Peterson
First Guide

to

MAMMALS

of North America

Peter Alden

Illustrated by
Richard P. Grossenheider

based on
*A Field Guide to the Mammals
of America North of Mexico*
by William H. Burt and
Richard P. Grossenheider

HOUGHTON MIFFLIN COMPANY

Boston   New York

For information about permission to reproduce selections from this book, write to Permissions, Houghton Mifflin Company, 215 Park Avenue South, New York, New York 10003

PETERSON FIRST GUIDES,
PETERSON FIELD GUIDES and
PETERSON FIELD GUIDE SERIES
are registered trademarks of
Houghton Mifflin Company.

Selected illustrations reproduced from *A Field Guide to the Mammals*, 3rd ed., copyright © 1952, 1964, 1976 by William Henry Burt and the estate of Richard Philip Grossenheider. Copyright © renewed 1980 by William H. Burt.

Library of Congress Cataloging-in-Publication Data

Alden, Peter.

Peterson first guide to mammals of North America.
Cover title: Peterson first guides. Mammals.
"Based on a field guide to the mammals of America north of Mexico, by William H. Burt and Richard P. Grossenheider."
Includes index.
1. Mammals—United States—Identification. 2. Mammals—Canada—Identification. I. Grossenheider, Richard Philip. II. Burt, William Henry, 1903–
. Field guide to the mammals. III. Title, IV. Title: First guide to the mammals of North America. V. Title: Peterson first guides. Mammals.
QL715.A43 1987 599.0973 86-27821
ISBN 0-395-91181-8

Printed in Italy

NIL 14 13 12

Editor's Note

In 1934, my *Field Guide to the Birds* first saw the light of day. This book was designed so that live birds could be readily identified at a distance, by their patterns, shapes, and field marks, without resorting to the technical points specialists use to name species in the hand or in the specimen tray. The book introduced the "Peterson System," as it is now called, a visual system based on patternistic drawings with arrows to pinpoint the key field marks. The system is now used throughout the Peterson Field Guide Series, which has grown to over thirty volumes on a wide range of subjects, from ferns to fishes, rocks to stars, animal tracks to edible plants.

Even though Peterson Field Guides are intended for the novice as well as the expert, there are still many beginners who would like something simpler to start with—a smaller guide that would give them confidence. It is for this audience—those who perhaps recognize a crow or robin, buttercup or daisy, but little else—that the Peterson First Guides have been created. They offer a selection of the animals and plants you are most likely to see during your first forays afield. By narrowing the choices—and using the Peterson System— they make identification much easier. First Guides make it easy to get started in the field, and easy to graduate to the full-fledged Peterson Field Guides.

Roger Tory Peterson

Introducing the Mammals

North America is rich with many different mammals living in many different habitats—forests, farms, prairies, deserts, swamps, mountains, oceans, and shores.

Many mammals are secretive and active only at night, so finding these species will take a great deal of patience, skill, and luck. But other mammals are active by day and are not at all shy. Some of the larger ones can best be seen in national parks and wildlife refuges. Many others can be seen—if you're alert and careful—in areas close to home in every state and province.

Mammal tracks are often easier to find than the animals themselves. Look for tracks in mud, sand, and snow and match them with the ones shown in this book.

Mammals are different from birds, reptiles, amphibians, and fish in a few important ways. Most mammals are covered with hair or fur, and almost all mammals have hair somewhere on their bodies. All mammals (with the exception of 3 very unusual species) are born live rather than hatched from eggs. All are warm-blooded, which means their body temperatures stay pretty much the same, regardless of the temperatures around them. And all mammals drink milk from their mothers' mammary glands when young.

Mammals are divided into smaller groups, depending on characteristics they have in common. The largest groups are called *orders*. The animals within each order are divided into smaller groups called *families*. For example, all mammals with special teeth for cutting meat belong to the Carnivore order. And the carnivores are divided into the Dog, Cat, Bear, Weasel, and Raccoon families.

The important characteristics of each order of mammals are described on the next few pages. Characteristics of some families are given throughout the book.

Note: Throughout the book, length is measured from the tip of the nose to the tip of the tail. For hoofed mammals (pp. 100–113), height is given instead, and is measured to the top of the shoulder.

MARSUPIALS p. 10

Marsupials have been on Earth longer than almost any other order. Their pea-sized young are born blind and naked and complete their development in a fur-lined pouch called a marsupium on the belly of their mother. While most marsupials are found in Australia and nearby islands, there are dozens of species in the Americas. One opossum species lives in N. America.

INSECTIVORES pp. 12–15

These small, energetic mammals have long, pointed snouts and tiny, beadlike eyes. Unlike mice, most insectivores (such as shrews and moles) have short dense fur covering their ears and have 5 clawed toes on each foot. Using their numerous sharp teeth, insectivores prey on insects, earthworms, spiders, fish, frogs, and carrion (dead animals).

BATS pp. 16–20

These are the only mammals that have wings and truly fly. (Flying squirrels and tropical mammals such as flying lemurs just glide.) A bat's wing is a tough membrane of skin that covers extra-long arm, hand, and finger bones. The membrane stretches from the forelimb, down the side of the body, to the leg. All North American bats use a form of "radar" called *echolocation* to catch their insect prey and to avoid objects. They give off high-pitched squeaks that strike objects and bounce back to their ears as echos. By listening to these echos, the bats can "read" their surroundings. Some bats roost in vast colonies in caves; others roost alone in trees or in attics, always hanging upside down. Some bats migrate south for the winter; others hibernate.

Bats are a major natural weapon against flying insect pests. Bat houses can be provided to encourage roosting in your backyard.

CARNIVORES

pp. 22–51

A pet dog or cat can introduce you to the basic features of carnivores. Carnivores have long canine teeth, sharp "cheek" teeth for slicing meat, 5 toes on the front feet, and 4 or 5 on the hind feet. While some eat nothing but animals they have killed themselves, others eat a good amount of plant food as well as meat. North America is home to both the world's smallest and largest carnivores: the Least Weasel, at $1/10$ of a pound, and the Alaskan Brown Bear, which can reach over 1500 pounds.

SEALS

pp. 52–57

Unlike whales and the other marine mammals on pp. 112–126, seals are closely related to carnivores. These ocean-dwelling mammals feed mostly on fish and other small aquatic animals. Their torpedo-like bodies feature front and hind flippers, and no tail is visible. Their nostrils close when submerged. Fine fur, which appears dark when wet but becomes paler when dry, covers most species of seals. Walruses have only a few sparse hairs. One seal—the Weddell—can dive to over 2000 feet and stay down for over 40 minutes. Seals and walruses "haul out" onto rocks, beaches, and ice floes only to rest, mate, and raise young.

RODENTS
pp. 58–91

These gnawing mammals outnumber those of all other orders combined, in both numbers of species and individuals. Rodents are active mostly at night and have bulbous eyes on the sides of the head. This helps them detect danger from nearly all directions at once. They are very active, and their high birth rate usually makes up for the great number of rodents that are caught by many predators. Most rodents have 4 toes on the forefeet (insectivores have 5) and 5 on the hind feet. They have 2 incisor teeth on top (rabbits have 4) and 2 below. Rodents lack canine teeth; they have a gap between the incisors and the molars.

RABBITS
pp. 92–99

Rabbits look a lot like rodents, but they have 4 upper incisor teeth rather than 2. There are 2 families in North America: the hares and rabbits in one and the pikas in another.

HOOFED MAMMALS
pp. 100–113

These heavy, plant-eating mammals have 2 toes on each foot. We have 4 families of hoofed mammals: the deer, the bovids (cattle), the pronghorns, and the peccaries. The first 3 families are *ruminants*. Their stomachs harbor bacteria and protozoa that aid in the digestion of plant fiber—a large part of these mammals' diets.

Harvest Mouse
and nest

EDENTATES pp. 112–113

These mammals live only in the Americas.
This group includes the anteaters, sloths,
and armadillos. One armadillo species lives
in North America, while all other endentates
live in Latin America. Anteaters have no teeth,
but the others have many peglike teeth.

SIRENIANS pp. 112–113

These large, nearly hairless, cylinder-
shaped mammals live in warm rivers and
coastal waters of the tropics and subtropics.
Their forelegs are in the form of flippers,
their hind legs are absent, and their tail is
broad and flat. Like whales, sirenians never
leave the water. But unlike whales, they eat
nothing but plants. The Manatee is the only
North American sirenian.

WHALES pp. 114–126

Although fishlike in form, whales (including
dolphins and porpoises) are true mammals.
All whales have a pair of front flippers and a
flat tail. They breathe through 1 or 2 nostrils on the top of the head, which are called
blowholes. The cloud of vapor created as
they exhale is called a blow, or spout, and is
useful in spotting and identifying different
species of whales. Most whales are fast
swimmers and deep divers, reaching up to
23 mph and remaining submerged up to 2
hours.

Within these groups (orders), the
sequence of species described and the
names used are essentially the same as in *A
Field Guide to the Mammals of America
North of Mexico*. Although we have included
all the more common and interesting North
American mammals in this First Guide,
these species are just a sampling. You will
soon be ready for the full treatment found
in that Field Guide, by William H. Burt and
Richard P. Grossenheider.

MARSUPIALS: OPOSSUMS

OPOSSUM   To 40 in. long
                               (including tail)*

This housecat-sized, nocturnal mammal is
whitish gray, with a *white face* and
pointed nose. Some Opossums (particularly
in the South) are blackish. All have thin,
round *black ears*. An Opossum can hang
from branches with its long, *rat-like tail*,
which is *pink* with a *black base*. The hind
feet have grasping "thumbs," which help
the Opossum grip branches and other
objects. Found on farms, in forests, and by
streams, the Opossum feeds on fruits, nuts,
bird eggs, insects, and carrion (dead animals). When frightened, an Opossum may
go into a state of shock and become so stiff
it appears to be dead. This behavior often
protects it from predators that attack only
live animals. Opossums are slowly expand-
ing their range from the Southeast and are
now also found throughout New England
and the Great Lakes region and along the
Pacific Coast. The dozen tiny young—each
the size of a lima bean—climb the mother's
long fur to her pouch immediately after
birth and stay there for another 2 months.

* length includes tail throughout book

OPOSSUM

r.f.

r.h.

2 in.

tail mark often seen

INSECTIVORES: SHREWS

Shrew Family

Shrews look like moles but are slimmer and have visible eyes. They feed under leaves on the ground, rarely burrowing into the ground. Because they are extremely active day and night, they must eat more than their own weight in food daily. There are 30 species in North America.

LEAST SHREW　　　　　　　To 3¼ in. long
Tiny and *cinnamon-colored*, with a very *short* (¾-in.) *tail*. Found throughout southeastern U.S., from Florida north to the Great Lakes and west to Texas and the Great Plains, chiefly in meadows and marshes.

MASKED SHREW　　　　　　　To 4½ in. long
Grayish brown, with a *long, bicolored tail* (brown above, *buff* below). Often the most common shrew in moist forests and brush in the northern U.S. and southern Canada.

ARCTIC SHREW　　　　　　　To 4⅔ in. long
Lives in bogs and among the leaf litter of northern forests. Has a *dark brown back* (nearly black in winter), *rusty sides*, and a *whitish belly*.

SHORTTAIL SHREW　　　　　　To 5⅕ in. long
Large and *lead-colored*. It has *no visible ears* and a *short tail*. Attacks worms, snails, and even mice, killing them with its poisonous saliva. Found in a wide range of habitats east of the Great Plains.

NORTHERN WATER SHREW　　To 6½ in. long
Large and *blackish gray*, with contrasting *silver underparts*. Can dive deeply and even run on the water after its prey. Found in the forests of New England, the Appalachians, and the Rockies north into Canada.

LEAST SHREW

MASKED SHREW

ARCTIC SHREW

SHORTTAIL SHREW

Shorttail Shrew foraging

NORTHERN WATER SHREW

INSECTIVORES: MOLES

Mole Family

Moles are stouter than shrews, with a *flexible, naked snout.* Their eyes are weak, but their senses of smell and touch are sharp. Moles have spade-like feet with *soles that turn outward.*

Low ridges meandering across lawns are a sign that moles have been moving just below the surface of the ground. They also "dive" deep into soft soils with a sort of breaststroke, sometimes digging at a speed of a foot per minute. They eat huge quantities of earthworms, grubs, slugs, insects, and occasional mice and tubers. The 7 North American species are absent from the Rockies and the Great Basin, where soils are too hard.

STARNOSE MOLE   To 8½ in. long
Dark brown to black, with a *star-shaped nose* of 22 tentacles. The hairy tail narrows near the body. Often seen above ground or swimming. It ranges from the Carolinas to Labrador and west to Minnesota.

HAIRYTAIL MOLE   To 7 in. long
A *slate-colored* mole with a *short, hairy tail.* Lives throughout the Appalachians, New England, and southern Canada.

EASTERN MOLE   To 8 in. long
The largest mole in the East and Great Plains. It has a *pink, naked, short tail.* The body varies from *slate-colored* in the North to *brown and golden* in the South and West.

TOWNSEND MOLE   To 9 in. long
Found only in the humid belt of northern California and the Pacific Northwest in moist meadows, gardens, and coniferous forests. As in the Eastern Mole, the front feet are *broader than long.* The tail is *slightly hairy.*

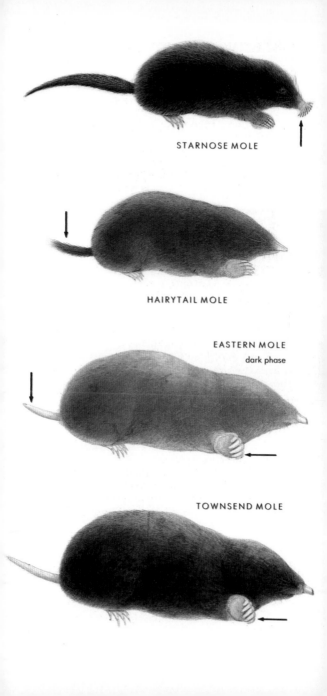

STARNOSE MOLE

HAIRYTAIL MOLE

EASTERN MOLE
dark phase

TOWNSEND MOLE

BATS

LEAFNOSE BAT   To 3½ in. long
A grayish bat with long ears. Like most
members of its family, it has a *leaflike flap*
of thick skin projecting upward from its
nose. It roosts in caves and old mine tunnels in Arizona, southern California, and
southern Nevada. Unlike most bats, it can
hover and also swoop down and snatch
large insects from the ground.

LONG-EARED MYOTIS   To 5½ in. long
Lives only in western coniferous forests,
often near buildings. Has *large black ears*
and long, glossy, *pale brown fur.*

LITTLE BROWN MYOTIS   To 5¼ in. long
Abundant in large colonies in all states and
provinces, but not found along Gulf Coast.
This bat is small and has medium-sized
ears. The black hair has long, *glossy tips.*
You may see it before sundown and recognize it by its erratic flight. Individuals that
breed in the North migrate south for the
winter.

BIG BROWN BAT   To 7 in. long
This common and widespread bat is large
and has *all-black wings.* A fast flier (up to
40 mph), it feeds mainly on beetles. It is
dark brown in moist areas and *paler brown*
in deserts. This bat often winters in buildings and is usually solitary.

LEAFNOSE BAT

LONG-EARED MYOTIS

LITTLE BROWN MYOTIS

BIG BROWN BAT

BATS

SILVER-HAIRED BAT   To 6 in. long
This bat is nearly black, but *silver-tipped hairs* on the neck give it a *frosted* look. Flies high and straight among trees in the middle and northern U.S. and Canada. Migrates south for the winter.

WESTERN PIPISTREL   To 5 in. long
Our smallest bat. It flies slowly (no more than 6 mph) and erratically in arid areas of the West and Southwest, often during the day. It is *pale yellowish* or ashy gray.

EASTERN PIPISTREL   To 5¼ in. long
One of the smallest eastern bats. It is *yellowish brown* to drab brown. It has a *slow, erratic flight* and is often on the wing before sundown.

RED BAT   To 7¼ in. long
The sexes of this bat are of a different color: males are *bright orange-red*, while females are *dull red*. Both have a *frosted look*, with hairs tipped with white. They roost alone on tree branches and feed in pairs while flying steadily and rapidly on a regular 100-yard course. They are widespread except in deserts and the Rockies, and they migrate south for the winter.

SEMINOLE BAT   To 6¼ in. long
This *mahogany brown* bat is found in the southeastern U.S. from Pennsylvania to Texas. It roosts low in trees and has habits similar to those of the Red Bat.

HOARY BAT   To 8½ in. long
This rarely seen bat lives throughout most of the U.S. and Canada and is the largest bat in the East. *White-tipped hairs* cover its *yellowish brown* to *mahogany brown* body, while the *throat is buffy*. Leaves late from its daytime roost in conifers and flies high. Migrates south for the winter.

SILVER-HAIRED BAT

WESTERN PIPISTREL

EASTERN PIPISTREL

female

male

RED BAT

SEMINOLE BAT

HOARY BAT

BATS

EASTERN YELLOW BAT　　　　To 6½ in. long
A *pale yellowish brown* bat with long, silky fur. It can be found in wooded areas of the southeastern coastal states from Virginia to Texas. Feeds mainly at medium altitudes and often roosts in clumps of Spanish moss.

EASTERN BIG-EARED BAT　　　To 6¼ in. long
This *pale brown* bat has *huge ears* joined in the middle and *2 prominent lumps on the nose.* It occurs throughout the Southeast and often hovers to pluck insects from plants.

PALLID BAT　　　　　　　　　To 7 in. long
This bat lives in the western deserts. It is the palest of those bats with *large ears.* When disturbed, it gives off a skunk-like odor. It feeds low to the ground, often picking up beetles, scorpions, and grasshoppers.

MEXICAN FREETAIL BAT　　　To 6¼ in. long
This bat and the next species are members of a bat family with a *tail extending well beyond* the edge of the *tail membrane.* This *chocolate brown* bat is our smallest "freetail" and is one of the commonest mammals in the southern U.S., with a population of more than 100 million. During the day it roosts in buildings and caves. In Texas and at Carlsbad, New Mexico, vast clouds of these bats exit from caves at dusk. They return at dawn, after feeding up to 150 miles away on moths and other insects.

WESTERN MASTIFF BAT　　　To 10¾ in. long
The *largest* bat in North America. It is *chocolate brown* with a *long, free tail* and *enormous ears.* It is found in small colonies in canyons and on cliffs along the Mexican border and in southern California. Its loud voice can be heard by humans.

EASTERN
YELLOW BAT

EASTERN
BIG-EARED BAT

PALLID BAT

MEXICAN
FREETAIL BAT

WESTERN
MASTIFF BAT

CARNIVORES: BEARS

Bear Family

Bears are the world's largest land-dwelling carnivores. They walk on the *entire foot*, rather than on the toes as cats and dogs do. They have short tails (usually not visible), small ears, and small eyes. Their eyesight is poor, but they have an excellent sense of smell. Most bears spend all winter in a den, where the female gives birth to her tiny cubs. Bears do not hibernate, but rather fall into a deep sleep from which they can awaken quickly.

BLACK BEAR   To 6 ft. long

Our *smallest* bear. In the East it is *nearly black* and in the West it is *black to cinnamon* or even *yellowish*. Its face is *roundish in profile* and always brown, and it usually has a white patch on its chest. Widespread in many forests, swamps, and mountain areas and is increasing in some parts of the East close to cities. Chiefly nocturnal, but it does forage in daytime, particularly in national parks. Eats berries, nuts, tubers, insects, small mammals, bird eggs, honey, carrion, and garbage. It can run up to 30 mph, and it weighs up to 475 pounds.

GRIZZLY BEAR   To 7 ft. long

Color varies from *pale yellowish* to *dark brown. White tips on its hairs* give it a grizzled appearance. Its face is *dish shaped* in profile, and it has a *distinct hump* at the shoulder, which is lacking in the Black Bear. Its front claws, which are longer than a Black Bear's, are useful in digging up rodents and excavating dens. Eats meat, fruit, grasses, fish, and other foods. Makes its own trails, which it uses over and over. At one time it was widespread in the Great Plains and the entire West. Now it survives mainly in national parks such as Yellowstone, Glacier, Banff, Jasper, and Denali. It can weigh up to 850 pounds.

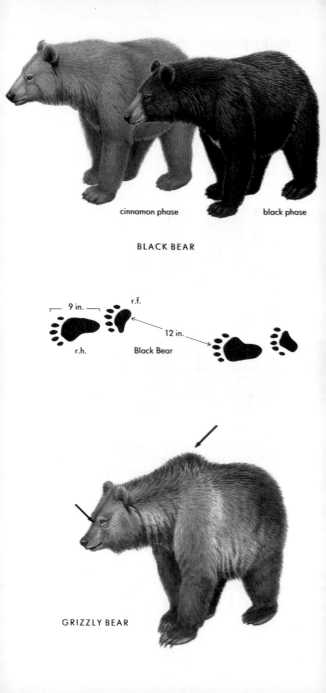

cinnamon phase

black phase

BLACK BEAR

9 in. — r.f.

12 in.

r.h. Black Bear

GRIZZLY BEAR

CARNIVORES: BEARS

POLAR BEAR   To 8¾ ft. long

A *very large, white* bear, often with a yellowish tinge. Its *nose is black* and the eyes reflect a silvery blue. It lives on ice floes, barren shores, and tundra around Hudson's Bay, throughout the Canadian Arctic, and along the coast of Alaska south to Nome. A strong swimmer, it often is seen in open water. In late summer and autumn it is common around the town of Churchill, Manitoba, where it has become a tourist attraction. This bear hunts by sneaking up on seals resting on ice or by waiting for them at their breathing holes in the ice. It also feeds on dead whales, birds, bird eggs, and some vegetation, such as mushrooms, grasses, berries, and seaweed. The polar bear dens for winter in steep snowbanks and emerges in late March. Males are quite a bit larger than females and can weigh up to 1100 pounds.

ALASKAN BROWN BEAR   To 9 ft. long

This bear—the world's largest—is closely related to the Grizzly. It shares the Grizzly's *dish-faced profile* and the noticeable hump above the shoulder, but its *front claws are shorter*, and it is usually *darker brown*, with a yellowish tinge. The Alaskan Brown Bear ranges along the mountains, coasts, and islands of southern Alaska. Those on Kodiak Island are particularly large. This bear emerges from its den in spring to feed on seaweed and carrion, foraging both day and night. As summer comes, it grazes on grasses and sedges and then congregates on rivers during salmon runs. It eats many rodents and stranded whales. Unprovoked attacks on humans are rare.

POLAR BEAR

ALASKAN BROWN BEAR

CARNIVORES: RACCOONS

Raccoon Family

Raccoons are dog-sized mammals with long-ish tails that feature rings or bands. They eat a wide variety of animal and vegetable foods and often live in small family groups.

RACCOON To 40 in. long

Has a *black "bandit" mask* that shows up well on its whitish face. The body is brownish gray and the *tail has rings* of black and yellowish white. Mainly nocturnal, but will forage in daylight. Eats fruit, nuts, grain, insects, bird eggs, and many aquatic animals such as frogs, salamanders, crayfish, and fish. Found mainly in forests along streams and is increasing rapidly in urban areas, where it raids garbage cans. Does not hibernate, but will den up in very cold weather. Found throughout the U.S. and southern Canada, except for some places in the Rockies and deserts.

COATI To 50 in. long

A mostly tropical mammal, found in the U.S. only in southeastern Arizona, southwestern New Mexico, and along the Rio Grande in Texas—mostly in forested hill country. Has a *long snout* and a *long, weakly banded tail* that is often *carried high*. Active by day in bands of up to a dozen. May be seen on the ground or in trees. It roots grubs and tubers from the ground with the aid of its tough nose pad. Also eats fruit, nuts, lizards, scorpions, and tarantulas. Also called coatimundi.

RINGTAIL To 31 in. long

A nocturnal denizen of woods, brush, and rocky areas—particularly along streams—from Oregon to Texas and into the Southwest. It is pale yellowish gray, with *short legs* and a *long, bushy tail* that has many *black-and-white rings*. It eats mice, birds, insects, lizards, and some fruit.

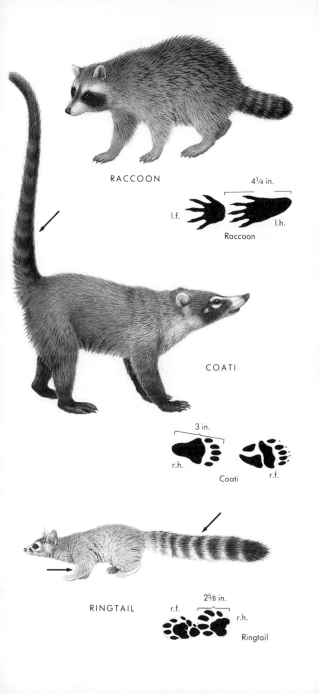

RACCOON

4¼ in.

l.f. l.h.

Raccoon

COATI

3 in.

r.h. Coati r.f.

RINGTAIL

2⅝ in.

r.f. r.h.

Ringtail

CARNIVORES: WEASELS

Weasel Family

This large family includes weasels, badgers, skunks, otters, and minks. All are small to medium-sized carnivores with low bodies; longish tails; short legs; short, rounded ears; and thick, silky fur. Many have strong scent glands used for defense, protecting food caches, and social signaling. Males often are much larger than females.

WOLVERINE   To 41 in. long

The largest member of the weasel family, found in the forests and tundra of northern and western Canada, Alaska, and in a few places in the Rockies and the Sierra Nevada. It weighs up to 60 pounds and looks like a small bear with a *heavy, bushy tail.* It has dark brown fur except for the yellowish temples and *broad yellow stripes* running from each shoulder to the base of the tail. This powerful and ferocious hunter preys on beavers, deer, porcupines, birds, and squirrels. It can kill prey as large as a moose or elk when the prey is bogged down in deep snow. A wolverine can drive bears and cougars from their kills and has a reputation for robbing traps and food caches of trappers. It marks any surplus supply of food with a smelly scent called musk, which repels other animals.

MINK   To 26 in. long

Found throughout most of Canada and the U.S., except for dry areas of the Southwest. Prowls the shores of ponds, lakes, and streams, hunting small mammals, birds, fish, crayfish, and frogs. The mink is an *excellent swimmer* and will pursue prey in water. It is a *rich dark brown* with a *white chin patch* and has a slightly bushy tail. Since it is mostly active at night, it is not often seen. Its eyes glow yellowish green when light is shined on them. Males can weigh up to 3 pounds; females are smaller.

WOLVERINE

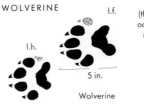

l.f.

(thumb print
occasionally
registers)

l.h.

5 in.

Wolverine

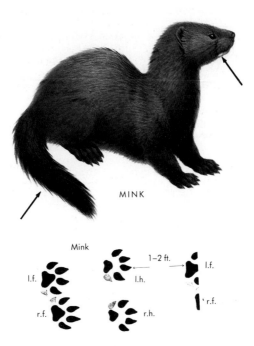

MINK

Mink

l.f.

1–2 ft.

l.h.

l.f.

r.f.

r.h.

r.f.

CARNIVORES: WEASELS

MARTEN   To 26 in. long

About the size of a Mink, but with *yellowish brown* fur and a bushy tail. Its legs and tail are darker brown, and it has a *pale, buffy patch* on its throat and breast. Found in the northern forests of Canada, Alaska, the Rockies, the Pacific Northwest, the upper Great Lakes, the Adirondacks, and northern New England. The Marten is a quick and agile tree-climber, and it specializes in capturing squirrels—particularly Red Squirrels. It also hunts rabbits, mice, and birds and feeds on bird eggs, berries, seeds, and honey when it finds them. It is chiefly nocturnal, but can also be seen early and late in the day. Fur trapping and lumbering have wiped out the Marten in many areas, but it is now making a comeback.

FISHER   To 40 in. long

The Fisher, weighing up to 12 pounds, is much larger than a Mink or Marten, and it lacks their pale chin and breast patches. It is *dark brownish black* overall, but white-tipped hairs give it a *frosted* appearance. The Fisher also is smaller and more slender than the Wolverine, with *no yellow stripes.* It lives in northern forests, sharing the range of the Marten, but it is absent from Alaska and the southern Rockies. It is active day and night both in trees and on the ground, but it is not as good a climber as the Marten. It preys on porcupines by flipping them over and attacking their undersides, which are unprotected by quills. It also feeds on snowshoe hares, squirrels, chipmunks, mice, fruit, and fern tips. Despite its name, the Fisher does not catch fish.

MARTEN

2 in. l.h.

Marten

l.f.

FISHER

3 in.

l.f.

Fisher

CARNIVORES: WEASELS

BADGER To 28 in. long

The Badger has a distinctive *black-and-white face* and a *white stripe* from its nose to its shoulders. The wide, flat body is yellowish gray, becoming more *yellowish* on the tail and belly. It has short black legs with extremely *long front claws*, which it uses for digging rodents from the ground. It feeds on ground squirrels, gophers, rats, mice, birds, and even rattlesnakes. Its long hair protects it from snakebites, unless a snake strikes it directly on the nose. The Badger forages day and night but is more nocturnal where it is threatened by humans. It can weigh up to 25 pounds and defends itself well when cornered.

BLACK-FOOTED FERRET To 24 in. long

At one time widespread in plains and high desert from Texas and Arizona north to Alberta, this ferret is now the most endangered mammal in North America. For millions of years it lived among the West's vast colonies of prairie dogs, feeding on the rodents as its main prey. But ranchers have poisoned almost every dogtown and thereby have driven the ferrets to near-extinction. At this writing a few pairs survive in captivity, and very small populations may still exist in Wyoming and elsewhere. This ferret has *yellowish buffy brown fur*, set off by *black feet* and a *black tail tip*. The face is white with a *black "bandit" mask*. In addition to feeding on prairie dogs, this ferret preys on ground squirrels, gophers, mice, birds, and small reptiles.

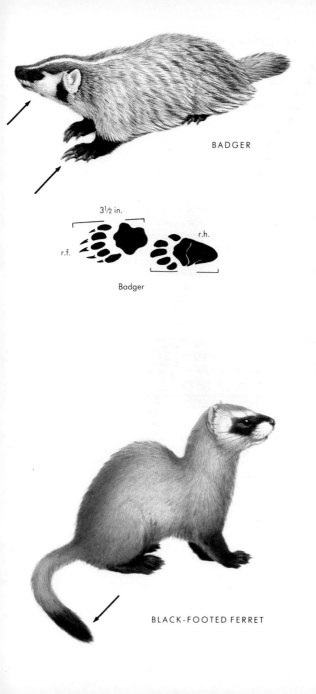

BADGER

3½ in.

r.f.

r.h.

Badger

BLACK-FOOTED FERRET

CARNIVORES: WEASELS

SHORTTAIL WEASEL To 13 in. long
(ERMINE)
In summer this weasel is dark brown above, with contrasting *white feet* and underparts and a white line down the inside of each hind leg. The Shorttail has a *black tip* on its tail even in winter, when the rest of its body becomes *pure white*. This winter color phase is called an Ermine. An expert mouser, the Shorttail also captures other small mammals and birds. It lives from Alaska and the Canadian Arctic south to Virginia, the Upper Midwest, the Rockies, and the Pacific Northwest, where it is light brown in winter rather than white. This weasel prefers bushy and wooded areas.

LEAST WEASEL To 8 in. long
This is the *smallest carnivore* in the world. It is all brown above and *white below, including its feet*. Its *short tail* is brown. In the North it is *all white in winter*. It ranges from Alaska and Canada south to the northern Great Plains, Great Lakes, and the Appalachians, but is not found in the West, the South, or New England. It prefers meadows and fields but also is found in open woods. It feeds at night, almost entirely on mice.

LONGTAIL WEASEL To 16½ in. long
The Longtail occurs in 48 states and southernmost Canada. It is the only weasel found in the Sun Belt. The Longtail superficially resembles the Shorttail in summer, but note the *longer, black-tipped tail*; the *yellowish white underparts*; and the *dark brown feet* (not white). Unlike the Shorttail, this weasel has no white line down the inside of each hind leg. In the North the Longtail becomes *white in winter* except for its black tail tip. In the Southwest a form with white patches on the face occurs. The Longtail feeds on mice, chipmunks, rats, and shrews, but can also catch prey larger than itself, such as rabbits.

34

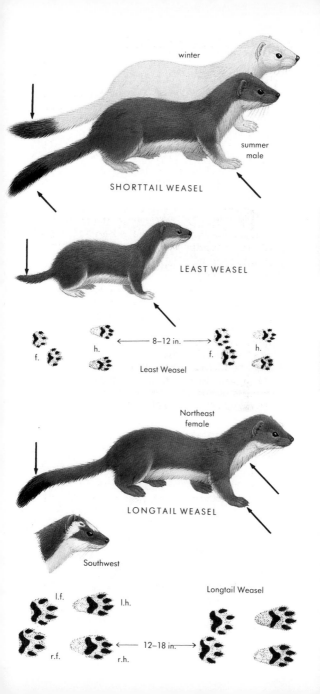

winter

summer male

SHORTTAIL WEASEL

LEAST WEASEL

f.   h.   ←— 8–12 in. —→   f.   h.

Least Weasel

Northeast female

LONGTAIL WEASEL

Southwest

Longtail Weasel

l.f.   l.h.

r.f.   r.h.   ←— 12–18 in. —→

CARNIVORES: SKUNKS

Skunks are black with various white patterns and have a long, fluffy tail, which they often hold high over the back. When threatened, skunks spray a stinging, acrid, yellowish liquid called musk from a gland at the base of the tail. A skunk can spray accurately at distances up to 15 feet. Skunks are omnivorous—they eat snakes, lizards, insects, frogs, fish, bird eggs, small mammals, fruit, corn, and carrion.

STRIPED SKUNK To 28 in. long

Our best-known skunk is found in 48 states and southern Canada. Note the white nape, the thin white stripe on its forehead, and the *2 broad white stripes* extending to its tail. Chiefly nocturnal, this skunk occurs in woods, prairies, suburbs, and even cities. It is active all year, but may den up in coldest weather.

SPOTTED SKUNK To 22 in. long

Our *smallest skunk*, with an irregular pattern of *white spots and stripes* all over its black body. It is sometimes found in forested areas, where it may climb trees. But it prefers open, brushy areas (including deserts) west of the Mississippi and south of the Ohio River.

HOODED SKUNK To 31 in. long

Found only along streams and on rocky ledges with thick brush, from southern Arizona and New Mexico to southwestern Texas. It has *2 thin white side stripes*, a shaggy *ruff on the neck*, and a *very long black tail*.

HOGNOSE SKUNK To 31 in. long

This skunk has a *long, pig-like snout*. The entire *back and tail are pure white* and the underparts are solid black. The Hognose Skunk is found from the Gulf Coast of Texas west to the Colorado River. It has *oversized front claws*, which it uses for digging up insects, rodents, mollusks, snakes, and tubers.

STRIPED SKUNK

SPOTTED SKUNK

HOODED SKUNK

HOGNOSE SKUNK

CARNIVORES: OTTERS

RIVER OTTER   To 47 in. long

A large, weasel-like aquatic mammal. *Rich brown above,* with a *silvery sheen below.* Has webbed feet and a long tail with a *thick base.* Swims rapidly under water or on the surface, stopping now and then to look around by raising its head. Feeds on fish, frogs, crayfish, and muskrats. The River Otter is a playful, sociable animal whose presence is often noted by slides worn into streambanks and snowbanks. It often runs several miles over land to visit distant lakes and streams. Originally found over most of the continent, the River Otter was eliminated from many areas by overtrapping and loss of habitat. It is now protected and is making a slow comeback in many areas. Weighs up to 25 pounds.

SEA OTTER   To 49 in. long

Found along the Pacific Coast from central California to the Aleutians of Alaska. It is best seen from Point Lobos (south of Carmel) in California and on Alaskan cruises. It feeds, sleeps, mates, and gives birth at sea, coming ashore only during severe storms. The Sea Otter swims on its back and often uses a rock as a tool to break sea urchins and abalones on its chest. Note its long whiskers; *huge webbed, flipper-like feet;* and *yellowish gray face and neck.* It is otherwise dark brown, with white-tipped hairs that give it a frosted look. The Sea Otter nearly became extinct by the early 1900s due to uncontrolled killing for its thick, valuable fur. With protection it is now making a comeback in some areas. This huge otter can weigh up to 85 pounds.

River Otter
slide on snow
or mud bank—
about 8 in. wide

l.h.

$2\frac{3}{4}$ in.

River Otter

l.f.

RIVER
OTTER

SEA OTTER

CARNIVORES: DOGS

Dog Family

The wolves, foxes, and coyotes have long, narrow muzzles; erect, triangular ears; long, slender legs; and bushy tails. Their sense of smell is excellent, and they have keen sight and hearing. Unlike cats, they are unable to retract their claws.

GRAY WOLF   To 67 in. long

Our largest and most powerful dog, with males weighing up to 120 pounds. This wolf usually is a *grizzled gray* with a black-tipped tail. Many individuals are nearly *white* (in the Arctic) or *nearly black. Carries its tail high* when running. Compared to the Coyote, a wolf has *short, rounded ears* and a broad face. It hunts both night and day in family groups (packs) of up to a dozen. A dominant male leads the pack, and the rest of the group help care for the young. Gray Wolves feed mainly on old, weak, and diseased deer, caribou, and moose, particularly those slowed by deep, crusty snow. They also feed on rodents, birds, fish, and berries. This wolf communicates by a wide variety of howls and barks. It once ranged throughout most of the continent, but now—due to relentless killing by humans—survives in good numbers only in Canada, Alaska, and Minnesota.

RED WOLF   To 57 in. long

A southern wolf with *reddish tawny legs, muzzle, and ears.* Weighing up to 70 pounds, it is smaller and less powerful than the Gray Wolf. The Red Wolf may be mistaken for the Coyote, but unlike a Coyote, the wolf runs with its *tail held high.* It once ranged from Pennsylvania to Texas, feeding chiefly on rodents, birds, and crabs. Shooting, trapping, habitat loss, and inbreeding with Coyotes have made it extinct in the wild. Captive-bred Red Wolves are now being reintroduced to a National Wildlife Refuge in North Carolina.

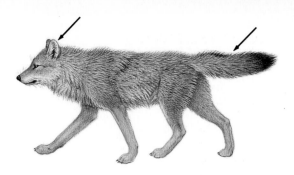

GRAY WOLF

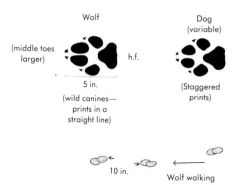

RED WOLF

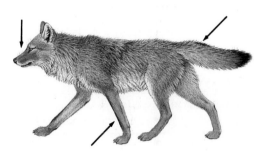

Wolf

(middle toes larger)

h.f.

5 in.

(wild canines—
prints in a
straight line)

Dog
(variable)

(Staggered
prints)

10 in.

Wolf walking

CARNIVORES: DOGS

COYOTE To 53 in. long

Larger than a fox but smaller than a wolf. It is gray or reddish gray, with *rusty legs, feet, and ears, and white underparts.* Unlike wolves, the Coyote runs with its *tail held down between its legs.* It also has *larger ears* than a wolf. The Coyote is active day and night and can be seen in most western wildlife refuges and parks. It usually hunts alone, for rabbits, hares, mice, ground squirrels, birds, frogs, and snakes. It can run up to 40 mph to catch faster prey. Occasionally, several Coyotes may down a larger animal such as a deer. Coyotes also feed on carrion (dead animals). Populations in the West are increasing, despite trapping, shooting, and poisoning by ranchers. Coyotes have adapted to suburbia in the West and have spread east to Massachusetts. A Coyote may weigh up to 50 pounds.

SWIFT FOX To 32 in. long

A rare fox of the Great Plains, from Alberta to Texas. *Buffy yellow,* with a conspicuous *black tip on its tail.* It is smaller than a Coyote, and it lacks the white-tipped tail and black lower legs of the Red Fox. It feeds on rodents and insects, but unfortunately has been eliminated from many areas by poisoning and habitat loss. The similar **Kit Fox** (not shown) lives in deserts west of the Rockies, from Oregon to Mexico.

ARCTIC FOX To 31 in. long

Found only beyond treeline in northern Canada and western Alaska. Like many other arctic mammals, it has *short, rounded ears* and heavily furred feet. Occurs in two color phases—the blue phase and white phase. In summer both types are brownish slate with no white tail tip. In winter some become *all white,* others *slate blue.* The Arctic Fox follows Polar Bears for scraps; scavenges on carcasses of marine mammals; captures lemmings, hares, and birds; and forages for berries.

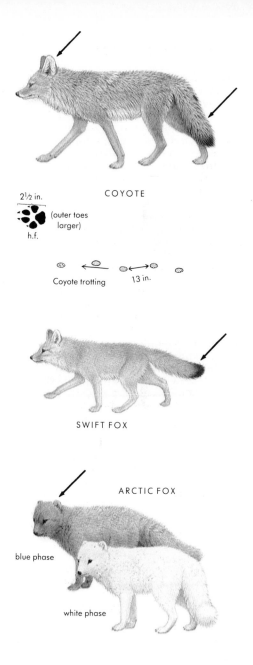

COYOTE

2½ in.

(outer toes larger)

h.f.

Coyote trotting   13 in.

SWIFT FOX

ARCTIC FOX

blue phase

white phase

CARNIVORES: DOGS

RED FOX   To 41 in. long
The classic quarry of the foxhunter, the wily
Red Fox is *reddish yellow* with *black legs.*
Its bushy tail is a mixture of black and reddish hairs; the *tail tip is white.* A rare color
variation is the black (or silver) phase,
which retains the white tail tip. The cross
phase resembles a Gray Fox but has a dark
cross over the shoulders and down the middle of the back. The Red Fox is active in
open country and forests almost everywhere, but is absent from much of the
Pacific Coast, southwestern deserts, and the
Rockies. It feeds on insects, birds, rodents,
rabbits, berries, and fruit—usually at
night. Weighs up to 15 pounds.

GRAY FOX   To 45 in. long
Note the *pepper-and-salt gray coat,* with
rusty patches on the neck, belly, legs, and
tail. The muzzle is blackish, contrasting
with the white cheeks and throat. The tail
does *not* have a white tip; a black stripe
down its entire length ends in a *black tip.*
The Gray Fox inhabits the eastern, southern, and southwestern U.S., barely reaching
Canada. It hunts in chaparral, open forests,
and rocky areas. It specializes in capturing
rodents, plus some insects, fruit, acorns,
birds, and eggs. This fox regularly forages
in trees, and will climb trees to escape enemies. The **Insular Gray Fox** (not shown)
inhabits 6 of the Channel Islands off the
coast of California.

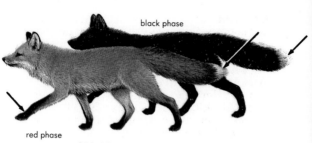

black phase

red phase

RED FOX

1¾ in.

h.f.

Red Fox

GRAY FOX

11 in.

Gray Fox trotting

CARNIVORES: CATS

Cat Family

Cats have shorter faces than most dogs do, with smaller ears. Unlike dogs, they have *retractile claws.* Their tails, whether short or long, are relatively thin, not bushy. Their vision is excellent. Most hunt at night.

MOUNTAIN LION   To 90 in. long

Many prefer to call this the Cougar or Puma. It is large and long, with a yellowish tawny or grayish coat. Its *long tail* is *tipped with dark brown,* as are the backs of the ears and sides of the face. While the young are spotted all over, the adults are unspotted. The Cougar survives mostly in forests, semi-arid areas, and mountains of the West. The only Cougars known east of the Mississippi live in southern Florida—a few individuals, known as the endangered "Florida Panther." The Cougar feeds mainly on deer, but it also hunts hares, rodents, Coyotes, Raccoons, and occasionally domestic animals. Chiefly nocturnal, it hunts by stalking its prey on the ground and by ambushing from trees. Males can weigh up to 200 pounds.

JAGUAR   To 84 in. long

This powerful cat ranged throughout much of the southern parts of California, Arizona, New Mexico, and Texas until the early 1900s. Its back and sides are tawny and evenly covered with *black spots* in the form of *rosettes* (circles of spots with a spot in the center). The belly is white with single black spots. The Jaguar is chiefly nocturnal and is rarely seen, even in Latin America. It hunts peccaries, rodents, and birds, and may kill cattle occasionally. Weighs up to 250 pounds.

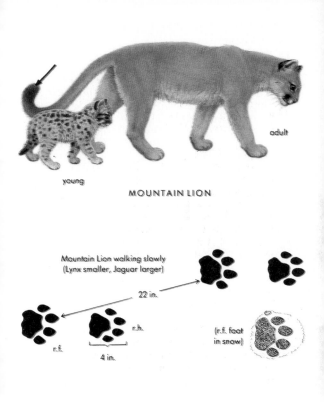

MOUNTAIN LION

young

adult

Mountain Lion walking slowly
(Lynx smaller, Jaguar larger)

22 in.

4 in.

r.f.

r.h.

(r.f. foot
in snow)

JAGUAR

CARNIVORES: TROPICAL CATS

OCELOT   To 50 in. long

The Ocelot has *rows* of blackish spots with dark centers, making it appear *striped*. It is much more heavily spotted than a Bobcat and has a much longer tail. The Ocelot is found in southern Texas and southeastern Arizona. It inhabits thick thornscrub, rocky areas, and woods. It hunts chiefly at night, and preys on rabbits, birds, fish, frogs, and rodents. It is adept at hunting in trees, in water, and on the ground. This cat suffered a huge decline in numbers due to the fur trade, pet trade, and loss of habitat. The sale of kittens and importation of pelts have been banned in the U.S. Weighs up to 40 pounds.

JAGUARUNDI   To 54 in. long

Another tropical cat, found primarily in Latin America but may be seen on rare occasions in southeastern Arizona and in southernmost Texas along the Rio Grande. Note its extremely *long, thin tail* and very short legs, similar to those of a weasel. This wild cat is twice as big as a house cat. It occurs in 2 color phases—all gray, or reddish. The Jaguarundi lives in bushy areas, thorn thickets, and mesquite. It is active in twilight and after dark. Feeds on rats, mice, rabbits, and birds. This cat swims well and also catches fish.

OCELOT

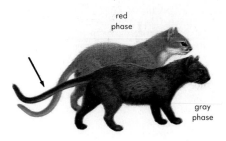

red
phase

gray
phase

JAGUARUNDI CAT

CARNIVORES: SHORT-TAILED CATS

LYNX   To 40 in. long

A *bob-tailed* cat of northern forests. Note the *long tufts* on the ears, the conspicuous facial ruff, and the *black tail tip*. The feet are *huge* and act as snowshoes, enabling this cat to walk over deep snow as it stalks and chases Snowshoe Hares, rodents, and birds. When populations of Snowshoe Hares crash—as they do naturally every 10 years or so—the Lynx also suffers a decline. It is found throughout much of Canada and Alaska and is also called the Canadian Lynx. The Lynx can also be seen in the upper Great Lakes region, Maine, the Rockies, and the Cascades. Trappers eagerly seek its luxurious, long, soft, grayish buff fur, which is mottled with brown. The Lynx is usually nocturnal, but it is forced to hunt by daylight in the far North during the long days of summer. Weighs up to 30 pounds.

BOBCAT   To 35 in. long

A southern cousin of the Lynx, the Bobcat is often called the Bay Lynx or "wildcat." In some areas its range overlaps that of the Lynx. The Bobcat can be recognized by its *dark spots* (particularly on the legs), its smaller ears, and the tip of its tail, which is black *only on top*. Its color varies from a warm tawny brown in summer to grayish in winter. This cat hunts birds and mammals, such as rabbits, hares, mice, squirrels, and porcupines, at night. It lives in forests, scrub, swamps, rocky country, and some farm lands, from the Canadian border southward to Mexico. The Bobcat has been heavily trapped for its fur, and is absent now in most of the Midwest. It is vocal at times, and can make a rich variety of coughs, yowls, and screams.

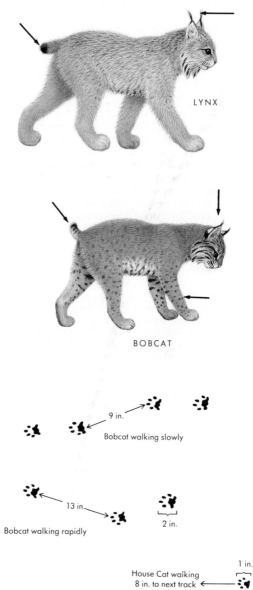

LYNX

BOBCAT

9 in.
Bobcat walking slowly

13 in.
Bobcat walking rapidly

2 in.

1 in.
House Cat walking
8 in. to next track

SEALS: EARED SEALS

Eared Seal Family

These seals have small *external ears*. Their hind flippers can be *turned forward*, which enables these seals to get about on land. Males are up to 4½ times larger than females.

NORTHERN SEA LION　　　To 10½ ft. long

An enormous sea lion—males weigh up to 2000 pounds, females up to 600 pounds. This sea lion has *yellowish brown fur* and a *low forehead*. Usually quiet when not molested, but it can give a lion-like roar. This seal lives on the *Pacific* Coast, from the Alaskan islands and bays south to California. It is also called Steller's Sea Lion.

CALIFORNIA SEA LION　　　To 8 ft. long

Smaller and darker than the Northern Sea Lion—males weigh up to 600 pounds, females up to 200 pounds. This sea lion has a *high forehead*. It is extremely noisy, barking constantly. This is the "seal" most often seen performing in aquarium shows and circus acts. Females and young remain along the California coast near their rookeries (breeding grounds), while males migrate northward, to British Columbia. This sea lion is a fast swimmer, capable of reaching 25 mph.

ALASKA FUR SEAL　　　To 6 ft. long

Males (which weigh up to 600 pounds) are *blackish* above, *reddish* on the belly, and have a *brownish face*. Females (which weigh up to 135 pounds) are gray above and reddish below. This fur seal breeds chiefly on the Pribilof Islands of Alaska and migrates as far south as California during the winter. Note the relatively *short face*. The **Guadalupe Fur Seal** (not shown), which has a *longer snout*, can be seen at San Nicolas Island, California.

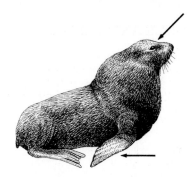

NORTHERN SEA LION
8–10½ ft.
(2.4–3.2 m)

**CALIFORNIA
SEA LION**
5½–8 ft.
(1.7–2.4 m)

**ALASKA
FUR SEAL**
4–6 ft.
(1.2–1.8 m)

WALRUS AND EARLESS SEALS

WALRUS To 12 ft. long
Like the eared seals, the walrus has hind feet (flippers) that can be turned forward for walking. However, like the earless seals, the Walrus has no external ears. Both males and females have *2 large white tusks* projecting downward from the upper jaw. The Walrus's thick, *hairless* skin appears black when wet, pinkish when dry. The Walrus dives down to the sea bottom to feed on mollusks and crustaceans, which it detects with its facial bristles. It sometimes catches small seals and will feed on dead whales. Found along coasts in the far North—from the Alaska peninsula to eastern Canada. Males weigh up to 2700 pounds.

Earless Seal Family
These seals have only an opening in the head for ears—*no visible ears.* Their hind flippers cannot be turned forward, so on land they are limited to wriggling. In all species except the Elephant Seal both sexes are roughly the *same size.*

ELEPHANT SEAL To 20 ft. long
The world's *largest* seal—males (bulls) weigh up to 8000 pounds (4 tons). Females weigh up to 2000 pounds (only 1 ton). Found along the Pacific Coast from California to the Gulf of Alaska. Most easily seen at Año Nuevo Point near Santa Cruz, California. This seal is essentially hairless, like a Walrus. Its color ranges from brown to grayish. Older males have *large, overhanging snouts,* which can be inflated to amplify their loud bellowing. The Elephant Seal sleeps on beaches by day and feeds on sharks, squid, and rays at night.

BEARDED SEAL To 10 ft. long
A large (up to 875 pounds), dark grayish to yellowish seal, with *long bristles* on each side of the muzzle. Found in arctic waters only as far south as western Alaska, Hudson's Bay, and Labrador. The **Gray Seal** (not shown), a large, grayish or black seal with a long face, replaces the Bearded Seal from Labrador south to Nantucket.

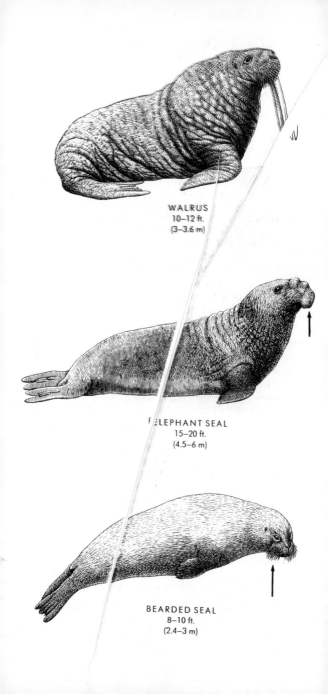

WALRUS
10–12 ft.
(3–3.6 m)

ELEPHANT SEAL
15–20 ft.
(4.5–6 m)

BEARDED SEAL
8–10 ft.
(2.4–3 m)

SEALS: EARLESS SEALS

HARBOR SEAL   To 5 ft. long
   The common small seal of harbors, river
mouths, and rocky coasts from the Carolinas and California north into the Arctic
and Hudson's Bay. Varies in color from
brown to gray, often with many *small spots.*
This seal hauls out on beaches and rocky
shores at low tide and feeds on fish, squid,
and octopus on the incoming tide. The
Ringed Seal (not shown) is similar but has
many *pale circles (rings)* on its dark back.
It is found on arctic coasts from Alaska to
Labrador.

RIBBON SEAL   To 5½ ft. long
   Inhabits ice floes and open waters of the
Bering Sea off western Alaska north of the
Aleutians. The males are brownish, with a
beautiful pattern of *wide, creamy or buff
rings* around the front flippers, neck, and
rump. Females are grayer, with indistinct
rings.

HARP SEAL   To 6 ft. long
   The male is a striking *yellowish white,*
with a *dark brown head* and variable *dark
patterns* on the back, with a few random
spots elsewhere. The female is less distinctly marked or has no dark markings.
The young are yellowish white all over, with
large, round black eyes. Harp Seals are
found from the Maritimes north to the eastern Canadian Arctic. They migrate up to
6000 miles a year to Baffin Island and the
Grand Banks of Newfoundland. The young
are born on the edge of the ice pack in the
Gulf of St. Lawrence. Each year, a limited
number of young are killed for their pelts.

HOODED SEAL   To 11 ft. long
   A large gray seal with numerous *large black
blotches.* The male has an inflatable "bag"
on top of his head which can be blown up to
make him look more formidable. Found
only in the northwest Atlantic, from the
Gulf of St. Lawrence to Baffin Island.

HARBOR SEAL
5 ft.
(1.5 m)

RIBBON SEAL
5 ft.
(1.5 m)

HARP SEAL
6 ft.
(1.8 m)

HOODED SEAL
7–11 ft.
(2.1–3.3 m)

LARGE RODENTS

BEAVER To 40 in. long

Our largest rodent (weighs up to 60 pounds). It has rich brown fur, huge gnawing teeth, webbed hind feet, and a naked, scaly, *paddle-shaped tail*. It fells trees by gnawing and uses branches to build stick and mud dams. The dams create protective ponds, in which it builds a domed lodge, or house. The Beaver may live along rivers, where it burrows into banks and does not build dams. It feeds on bark and small twigs, stashing a supply of branches underwater for winter use. At one time found almost everywhere in North America, the Beaver has survived heavy trapping and is being reintroduced into many areas, including western rangelands.

PORCUPINE To 31 in. long

Our only mammal with *long sharp quills,* which cover all but the belly. *Heavy-bodied* and *short-legged,* it is most often seen as a black ball resting high in a tree, where it eats bark, buds, and small twigs. Active chiefly at night. The Porcupine can cause damage to trees and wooden buildings and poles, especially in areas where its predators have been wiped out. It lives in forests of Alaska, Canada, the West, the upper Great Lakes, and the Northeast.

APLODONTIA To 18 in. long

The world's most primitive rodent. Looks like a plump, *tailless* Muskrat with small eyes and ears. Found mostly in the moist forests of the Pacific Coast, from San Francisco Bay and the Sierra Nevada north through western Oregon, Washington, and southern British Columbia. Makes extensive runways and burrows beneath dense streamside vegetation. The Aplodontia is active mostly at night, when it emerges to feed on herbaceous plants and on shrubs. Also known as Mountain Beaver.

Beaver

3–6 in.
About 4 in.
between tracks.
Hind track
covers front.

BEAVER

Tree cut by Beaver

PORCUPINE

r.f.

r.h.

3 in.

Porcupine

APLODONTIA

AQUATIC RODENTS

MUSKRAT   To 25 in. long
The Muskrat has dense, rich brown fur and
a silver belly. It features a long, scaly, naked
tail that is flattened from side to side
(rather than top to bottom, as in the Beaver, which is much larger). Muskrats use
aquatic plants in marshes to build conspicuous *conical houses* that rise up to 3 ft.
above the waterline. They also live in lakes
and streams, where they dig burrows with
underwater entrances into banks. A Muskrat marks its territory by leaving musky
secretions on vegetation. It feeds on cat-
tails, sedges, rushes, water lilies, frogs,
clams, and fish (rarely). It is often seen and
is frequently trapped for its valuable fur
over most of the U.S. and Canada except for
the far North, Florida, and most of the
Southwest. Weighs up to 4 pounds.

FLORIDA WATER RAT   To 15 in. long
Also known as the Round-tailed Muskrat,
this water rat *lacks the flattened tail* of the
northern Muskrat. It replaces its larger relative in Florida and the Okeefenokee area of
Georgia. The Florida Water Rat builds *bulky
nests* out of tightly woven sedges at bases of
stumps and mangroves as well as in savan-
nas near streams. It feeds on water plants
and crayfish in bogs, marshes, lakes, and
everglades. Weighs only 12 ounces.

NUTRIA   To 42 in. long
This large aquatic rodent weighs up to 25
pounds. A native of Argentina and Brazil, it
escaped from Louisiana fur farms in the
1940s and has replaced Muskrats in much
of the South and Pacific Northwest. It is
grayer than the Muskrat, with a *longer,
round tail*. It nests among marsh plants or
digs a burrow in streambanks above the
waterline.

Muskrat houses in marsh

MUSKRAT

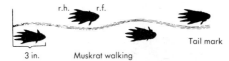

r.h.   r.f.

Tail mark

3 in.   Muskrat walking

FLORIDA WATER RAT

NUTRIA

RODENTS: MARMOTS

Marmots are *oversized squirrels*, with *short legs* and long digging claws on the front feet. All hibernate during winter.

WOODCHUCK (GROUNDHOG)　　To 27 in. long
This marmot is found throughout most of Canada and the northeastern and midwestern states. It is a uniform *frosted brown*, with *dark feet* and some white around the nose. It digs a den up to 5 ft. deep and up to 30 ft. long. Woodchucks feed on plants such as grasses, clover, and alfalfa. In some areas they are a nuisance in gardens and corn fields. Active day or night, they often can be seen swimming, resting in trees, and feeding beside roadways. The Woodchuck differs from the Beaver and Muskrat by its medium-length bushy tail. Its call is a shrill whistle. Weighs up to 10 pounds.

YELLOWBELLY MARMOT　　To 28 in. long
Replaces the Woodchuck in the interior western U.S., from New Mexico to British Columbia and from the Black Hills to California. It is a rich *yellowish brown*, with a yellow belly and bushy tail, which is reddish with a black tip. The *black head* has *white patches* in front of the eyes and a rusty patch below each ear. This marmot feeds on grasses and forbs in mountains and valleys, usually near rocky areas. It gives *high-pitched chirps* from a lookout boulder when alarmed.

HOARY MARMOT　　To 31 in. long
A very large marmot (weighs up to 20 pounds). *Silvery gray*, with a *black-and-white* head and shoulders. It inhabits rockslides near meadows in mountains of Alaska and western Canada south to Idaho. Gives a *shrill whistle*. Males may be seen engaged in stand-up wrestling matches. Three very similar marmots (not shown) are found only on Vancouver Island, the Olympic Peninsula, and northern Alaska.

Main entrance

Secret entrance

Woodchuck den

1½ in

l.h.   l.f.

WOODCHUCK

YELLOWBELLY
MARMOT

HOARY
MARMOT

RODENTS: PRAIRIE DOGS

The western plains at one time swarmed with billions of these chunky, broad-headed squirrels. Both species have very *short, hairy tails;* short legs; and strong claws. These social animals live in "towns" and post sentinels to warn the colony of incoming coyotes, badgers, ferrets, snakes, and birds of prey. They *bark* (like a dog) and bob up and down in excitement before retreating below. Their underground villages can be 16 ft. deep and can extend another 20 ft. on the level, with side chambers for storage and nesting and with escape tunnels. Prairie dogs feed on grasses, roots, and blossoms. In the past, ranchers have shot, poisoned, trapped, and gassed them to keep them from making new burrows, fearing that livestock would break a leg in the prairie dogs' entrance holes. Grain farmers of today have eliminated most of the rest. Those that survive are found in parks and other uncultivated areas.

BLACKTAIL PRAIRIE DOG   To 21 in. long
This prairie dog ranges across the *short-grass* prairies of the western Great Plains from Montana to western Texas. It weighs up to 3 pounds and can be identified by its *black-tipped tail.* A large colony can be seen at Wichita Mountain National Wildlife Refuge, in Oklahoma.

WHITETAIL PRAIRIE DOG   To 14½ in. long
Found in upland meadows and brushy country with scattered junipers and pines, at altitudes of 5,000–12,000 ft. Ranges farther west than the Blacktail, into Wyoming, Utah, Arizona, western Colorado, and western New Mexico. Has a *shorter tail* than the Blacktail, with a *white tip.*

BLACKTAIL PRAIRIE DOG

WHITETAIL PRAIRIE DOG

Prairie dog town

RODENTS: GROUND SQUIRRELS

A variable group of small to medium-sized, ground-dwelling squirrels found in the western half of the continent and on the tundra. Many sit up like prairie dogs, but ground squirrels have longer faces or longer tails, or both. Eight of the 17 species are shown here. (Consult a *Field Guide to the Mammals* for the details on the rest.)

CALIFORNIA GROUND SQUIRREL
To 20 in. long

Brown with *buffy bands* on rear half. Dark back contrasts with *pale gray neck* and shoulders. The white-fringed tail is much less bushy than that of the Western Gray Squirrel. This ground squirrel prefers pastures, grain fields, lightly treed slopes, and rocky ridges. It eats a wide variety of green vegetation, fruits, mushrooms, acorns, seeds, berries, bird eggs, and insects. Its burrows are up to *200 ft. long.* Found in more humid areas of California and western Oregon.

ROCK SQUIRREL
To 21 in. long

The *largest* ground squirrel of the Southwest, ranging from Great Salt Lake and the Colorado Rockies south to the Mexican border of Arizona and western Texas. A slightly *mottled gray* squirrel with a somewhat *bushy tail.* Some individuals are darker on the head and back. Inhabits rocky canyons and slopes with boulders.

COLUMBIAN GROUND SQUIRREL
To 17 in. long

Rather large, with a short, bushy tail, *mottled gray upperparts,* and *reddish legs* and *face.* Lives in meadows and at forest edges from the Canadian Rockies south to Idaho.

ARCTIC GROUND SQUIRREL
To 21 in. long

A large squirrel of Alaska and northern Canada, west of Hudson's Bay. Found in tundra and brushy mountain meadows. The only ground squirrel in its range. *Dusky brown* with many *pale spots.* Very vocal—sounds include a *sik-sik* call.

CALIFORNIA GROUND SQUIRREL

ROCK SQUIRREL

COLUMBIAN GROUND SQUIRREL

ARCTIC GROUND SQUIRREL

RODENTS: GROUND SQUIRRELS

THIRTEEN-LINED GROUND SQUIRREL   To 11½ in. long

Unique *dark brown stripes* above, with *buffy spots* within them. This ground squirrel originally lived in the shortgrass prairies from Alberta to Texas, but—with the clearing of forests—has spread eastward to Ohio and Michigan. It is often seen feeding on seeds and insects along roadsides and on lawns and golf courses.

SPOTTED GROUND SQUIRREL   To 9½ in. long

A pale brownish squirrel, with *light buffy spots* on its back. The tail is pencil-like, not bushy. A shy resident of sandy forests, brush, and grassy parks from Nevada to the Rio Grande and eastern Arizona. In southwestern Texas and southeastern New Mexico, look for the **Mexican Ground Squirrel** (not shown), which is similar but *darker brown,* with *whiter spots* in distinct *rows.*

FRANKLIN GROUND SQUIRREL   To 16 in. long

Found in the *Midwest* and northern Great Plains, from Missouri to Saskatchewan. The *largest* and *darkest* ground squirrel in its range. Dark gray with tawny overtones and a fairly long tail. Will climb trees, but usually seen on the ground. The **Uinta Ground Squirrel** (not shown) is similar (drab gray and brown), but has a *cinnamon-colored face.* It is seen near lodges at Yellowstone and Grand Teton.

GOLDEN-MANTLED SQUIRREL   To 12½ in. long

A ground squirrel that looks like a chipmunk. Familiar to campers and visitors in most western parks, as it becomes tame due to hand-outs. Note its *copper-red head* and *shoulders,* "chipmunk" stripes (only on the back), and relatively small tail. Found in evergreen forests, in chaparral, and above timberline.

THIRTEEN-LINED
GROUND
SQUIRREL

SPOTTED
GROUND
SQUIRREL

FRANKLIN
GROUND
SQUIRREL

GOLDEN MANTLED
SQUIRREL

RODENTS: CHIPMUNKS AND ANTELOPE SQUIRRELS

Chipmunks

Alert, small, ground-dwelling squirrels with a slightly bushy tail that is flicked often. Note their narrow, *erect ears* and *cheek pouches*, which are frequently stuffed with food. All 16 species feature 5 dark and 4 paler *stripes down the back* and more or less distinct *white stripes* above and below the eye. Chipmunks feed on seeds, nuts, fruit, insects, and bird eggs.

Chipmunks are territorial and often chase each other to defend their burrows or food caches. A burrow can be 30 ft. long with side chambers, escape hatches, and an entrance by a boulder or stump.

TOWNSEND CHIPMUNK To 12½ in. long
Large and dark brown. Found in the Pacific Northwest and Sierra Nevada.

EASTERN CHIPMUNK To 10 in. long
Reddish rump. Eastern U.S. and Canada.

MERRIAM CHIPMUNK To 12 in. long
Stripes are dark brown (not black) and gray. Central and southern California.

LEAST CHIPMUNK To 9 in. long
Smallest and most variable in color. Stripes extend to *base of tail.* Ontario to Yukon, and interior western U.S.

CLIFF CHIPMUNK To 10 in. long
Grayish with *indistinct stripes.* Southwest.

COLORADO CHIPMUNK To 9½ in. long
Large *white patch* behind each ear. Upper Colorado Basin and southern Rockies.

Antelope Squirrels

Four similar species live in the *deserts* of the Southwest and Great Basin. Each has a *white stripe* on each side of back. All run with the *tail arched over the back.* The **Yuma** of southern Arizona is replaced by the **Whitetail** (not shown) between the Rockies and Sierra Nevada.

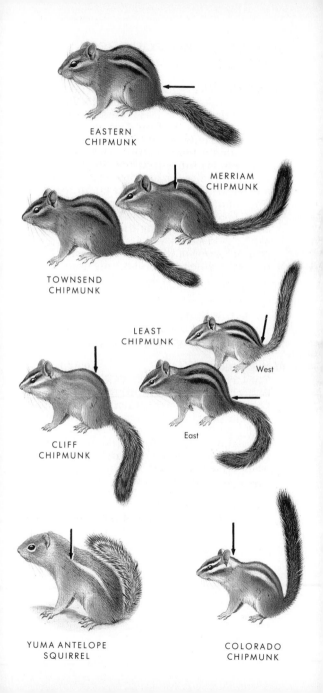

EASTERN
CHIPMUNK

MERRIAM
CHIPMUNK

TOWNSEND
CHIPMUNK

LEAST
CHIPMUNK

West

East

CLIFF
CHIPMUNK

YUMA ANTELOPE
SQUIRREL

COLORADO
CHIPMUNK

RODENTS: TREE SQUIRRELS (EASTERN)

EASTERN GRAY SQUIRREL To 20 in. long
Perhaps the most familiar mammal of eastern North America. It is abundant in city parks, suburbs, and rural woodlands where there are plenty of nut trees. Its tail is very bushy and is bordered with white hairs. The body is *gray* with white underparts, and in summer it may appear tawny. Black and all-white individuals are not unusual. This squirrel feeds on a great variety of nuts, seeds, fungi, and fruits. It can destroy small trees by stripping the bark to reach sap. It stores nuts and acorns in the ground, many of which sprout. Nests in natural cavities and builds leafy nests in tree branches. As it dashes among tree branches, its tail helps it balance. It becomes a pest by stealing seeds people put out to feed birds, and by gnawing its way into buildings to spend the winter. Barks sharply or chatters when excited.

FOX SQUIRREL To 29 in. long
The largest tree squirrel. Widespread in eastern U.S. and westward through the Great Plains, but absent northeast of Philadelphia. Normally it is *rusty* yellowish gray with a pale *yellowish orange belly*, but populations in the Chesapeake Bay region are often *pure gray* (no tawny) with white around the nose. In the Southeast it often has a *black head* with white on the nose and ears. Its habits, food, and calls are similar to those of the Eastern Gray Squirrel, but it can weigh twice as much (up to 3 pounds).

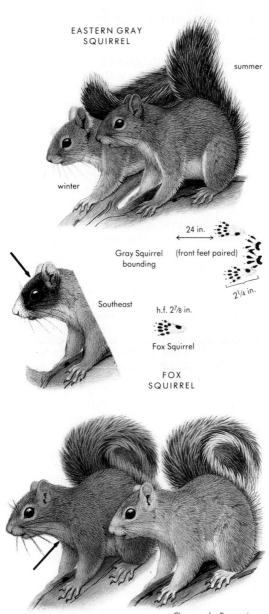

EASTERN GRAY
SQUIRREL

summer

winter

24 in.

Gray Squirrel
bounding

(front feet paired)

2¼ in.

Southeast

h.f. 2⅞ in.

Fox Squirrel

FOX
SQUIRREL

Chesapeake Bay region

RODENTS: TREE SQUIRRELS (WESTERN)

WESTERN GRAY SQUIRREL To 24 in. long
A large, gray squirrel with a long, bushy tail and white belly. Lives in California, western Oregon, and Washington, in oak and pine-oak forests. Differs from the Eastern Gray (which has been introduced near Seattle and Vancouver) by its *dusky feet*, the reddish fur on the back of the ears, light banding on the underside of the tail, and lack of tawny overtones in summer. Feeds on acorns and evergreen seeds and thrives on walnut and almond plantations.

TASSEL-EARED SQUIRREL To 21 in. long
America's most attractive and colorful tree squirrel. Gray, with a *chestnut back* and chestnut on the back of its *tasseled ears.* Note the contrast between *dark sides* and white underparts. The tail is usually gray above, *white below.* A tourist attraction in yellow pine forests of Arizona, New Mexico, Colorado, and southeastern Utah. Feeds on pine seeds, pinyon nuts, mistletoe, and inner tree bark. An isolated population lives on the north rim of the Grand Canyon. These individuals have an *all-white tail* and black (not white) underparts; they are known as the Kaibab Squirrel.

CHICKAREE To 12 in. long
Also known as the Douglas Squirrel. This small tree squirrel replaces the Red Squirrel (p. 76) in the humid evergreen forests of the Pacific Northwest, from coastal British Columbia south to the Sierra Nevada. The *tail is blacker* than that of the Red Squirrel. *Dark gray* in summer; in winter it is paler gray, with *gray ear tufts*. Its habits are similar to those of the Red Squirrel.

WESTERN GRAY
SQUIRREL

south of
Grand Canyon

TASSEL-EARED
SQUIRREL

north rim of Grand Canyon
(Kaibab Squirrel)

CHICKAREE

summer

winter

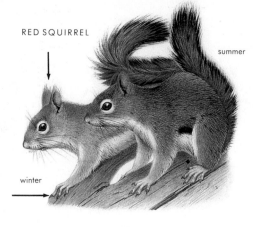

RED SQUIRREL

summer

winter

Red Squirrel
h.f. 1¾ in.

RODENTS: SMALL TREE SQUIRRELS

RED SQUIRREL To 14 in. long
Considerably smaller than the Eastern or
Western Gray Squirrel. This *rusty red*
squirrel has distinct summer and winter
coats, though it is always *white below*, with
a *red tail*. In summer it is a *darker* reddish
gray, with *black* side stripes. In winter it is
a *paler* rusty gray, with *rusty ear tufts* and
no side stripes. The Red Squirrel lives in
pine, spruce, and hardwood forests of
Alaska, Canada, the Rocky Mountains
(where it is called the Pine Squirrel), and
the northeastern U.S. south to the Smokies.
It feeds on seed, nuts, and fungi. Active all
year, and very vocal.

Flying Squirrels
Small, *nocturnal* squirrels with *large* black
eyes. *Gray-brown* to cinnamon above, white
below. These squirrels have broad folds of
skin that connect their front and back feet,
allowing them to glide down (not fly) from

Leaf nest of
Tree Squirrel

Flying Squirrel gliding
from den-tree hole

one tree to another. The flat, broad tail helps break their falls. Flying squirrels spend their daylight hours in natural and manmade cavities, emerging to feed on seeds, nuts, and insects. They utter faint, birdlike *chip* notes.

SOUTHERN FLYING SQUIRREL   To 10 in. long

Found in the eastern U.S. and southern Ontario. Olive-brown above and pure white below. The **Northern Flying Squirrel** (not shown) is slightly larger (up to 12 in. long). It is found in the western U.S. and much of Canada; its range overlaps that of the Southern Flying Squirrel in the Great Lakes and New England southward through the Appalachians.

SOUTHERN
FLYING
SQUIRREL

RODENTS: NATIVE MICE

WHITE-FOOTED MOUSE　　　　To 8 in. long
Widespread in forests and brushy areas
from New England to Arizona and Montana.
Note its *large ears* and rich reddish brown
upperparts, contrasting with the pure
white feet and underparts. The tail is *usually no longer than the body.* This mouse
lives in abandoned bird nests, outbuildings,
and other animals' burrows. Active year
'round, it feeds on seeds and insects.

DEER MOUSE　　　　To 9 in. long
Our most wide-ranging native mouse. Has a
bicolored tail and its body *color varies* from
grayish buff to deep reddish brown. Occurs
throughout Canada and U.S. except for
Alaska and the Southeast. The larger woodland form nests in hollow logs; the prairie
form digs small burrows. Feeds on seeds,
nuts, centipedes, and insects.

GOLDEN MOUSE　　　　To 7½ in. long
A handsome, *golden-cinnamon* mouse with
white belly. Lives only in the *southeastern*
U.S., among *trees,* vines and brush. This
mouse feeds on insects and seeds of poison
ivy, sumac, greenbrier, and wild cherry.

JUMPING MICE　　　　To 10 in. long
Four species of very *long-tailed* mice with
large hind feet and *small ears* edged with
white or buff. They are widespread except in
deserts and the Southeast. The **Woodland
Jumping Mouse** lives in the Northeast and
is active at night, while the **Meadow Jumping Mouse** (not shown) is often seen by day.

HARVEST MICE　　　　To 6¼ in. long
Five species of small brown mice that
resemble the House Mouse. Active day and
night, they live in dense vegetation and
build large grass nests (see p. 8). These
mice feed on vegetation, seeds, and fruit.
They are found in the southeastern and
western U.S.

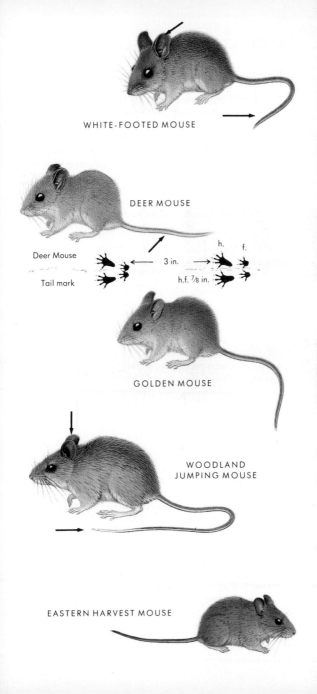

WHITE-FOOTED MOUSE

DEER MOUSE

Deer Mouse

Tail mark

h.   f.

← 3 in. →

h.f. ⁷⁄₈ in.

GOLDEN MOUSE

**WOODLAND
JUMPING MOUSE**

EASTERN HARVEST MOUSE

RODENTS: NATIVE MICE

GRASSHOPPER MICE To 7 in. long
Pale cinnamon or *pale gray* above, with white underparts and a *short white tail.* These stout, largely carnivorous mice prey on other mice, grasshoppers, scorpions, beetles, and lizards. Occasionally they eat seeds. These mice live in burrows of other animals and appear at night in open sandy or gravelly sage and grasslands. The **Northern Grasshopper Mouse** occupies the Great Plains, Great Basin, and western plateaus, while the smaller **Southern Grasshopper Mouse** (not shown) replaces it in southwestern deserts.

PINYON MOUSE To 9 in. long
This mouse has even *larger ears* than the Brush Mouse. The tail is bicolored and is often slightly *shorter* than the body. The Pinyon Mouse lives in rocky terrain with scattered pinyon pines and juniper.

BRUSH MOUSE To 8 in. long
A *large-eared,* gray-brown mouse, with tawny sides and a *well-haired tail* that is often slightly *longer* than the body. This mouse lives in chaparral and brush of semi-arid areas from Arkansas west to California.

HISPID COTTON RAT To 14 in. long
Widespread in the South, from the Carolinas to the Rio Grande. Note its *long, coarse fur,* heavily mixed with buff and black hairs, and its *tail,* which is *shorter* than the body. Lives in moist tall grass and feeds on vegetation and bird eggs.

RICE RAT To 12 in. long
Grayish brown above, with a pale gray or buff belly. The *scaly tail* is *longer* than the body and the feet are *whitish.* Inhabits marshy areas of southeastern states. Active chiefly at night, building or traveling on runways through grass. This rat feeds on seeds, vegetation, crabs, snails, and insects. It swims well.

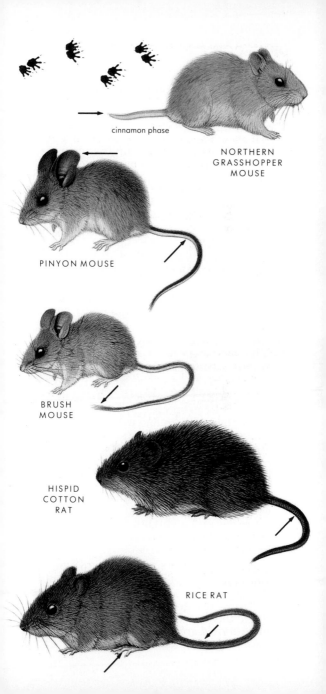

cinnamon phase

NORTHERN
GRASSHOPPER
MOUSE

PINYON MOUSE

BRUSH
MOUSE

HISPID
COTTON
RAT

RICE RAT

OPEN-COUNTRY RODENTS

Kangaroo Rats

Fourteen species of nocturnal rodents that live in arid areas of the western U.S. They have extremely *long, white hind feet,* and *powerful thighs* (crossed by a *white line*) that allow them to jump up to 6 ft. in one bound. The front feet are tiny by comparison. The dark tails with *white side stripes* and a *bushy tip* are much longer than the body. Size varies and body color varies from a pale sandy color to dark brown. Kangaroo rats feed on seeds and desert foliage.

BANNERTAIL   To 15 in. long
KANGAROO RAT

Boldly marked, with a *black-and-white tail.* Found in New Mexico, west Texas, and southern Arizona.

ORD KANGAROO RAT   To 10½ in. long

Great Plains and Great Basin from Canada to Mexico.

MERRIAM KANGAROO RAT   To 10 in. long

Our smallest species. California north to Nevada, east to Texas.

PLAINS POCKET GOPHER   To 13 in. long

One of a dozen species of pocket gophers living in the southern and western U.S. and Canadian prairie provinces. These rodents are named for their *fur-lined cheek pouches* (pockets). Note their large, yellowish, *exposed incisor teeth; heavily clawed front feet;* and a *short, naked, rat-like tail.* Pocket gophers make fan-shaped mounds when they dig their burrows. They are active day and night, but rarely leave their burrows. They feed on roots, tubers, and field crops.

CALIFORNIA POCKET MOUSE   To 9 in. long

One of the 20 species of small, nocturnal desert and plains mice with *cheek pouches* like those of a gopher and legs and feet similar to those of kangaroo rats. They are unpatterned gray or brownish above, and have long, thin tails.

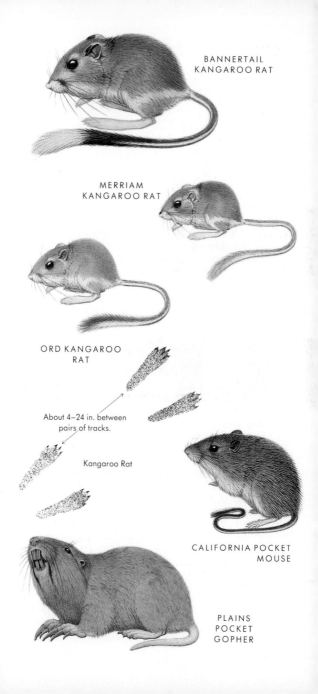

BANNERTAIL
KANGAROO RAT

MERRIAM
KANGAROO RAT

ORD KANGAROO
RAT

About 4–24 in. between
pairs of tracks.

Kangaroo Rat

CALIFORNIA POCKET
MOUSE

PLAINS
POCKET
GOPHER

RODENTS: VOLES

Redback Voles
BOREAL REDBACK VOLE   To 6½ in. long
A small, relatively tame inhabitant of cool,
moist forests and bogs. Uses natural runways along logs, rocks, and trails. This
vole's *reddish back* contrasts with its gray
sides, although those in the Northeast are
all gray. Widespread in northeastern U.S.,
upper Midwest, Rockies, and Canada. The
California Redback Vole (not shown) of the
Pacific Northwest has dark olive-brown
sides and a dark chestnut back.

Other Voles
Active day and night, these rodents of
grassy areas dig underground burrows and
make runways up to 2 in. wide through
matted grasses. In winter, the round
entrances of their tunnels can be seen in
snow. Voles feed mainly on plants and
sometimes eat seeds or bark from tree roots
as well as green vegetation. The 17 species
are mainly brownish gray with long fur.
They have *small ears,* a tail that is *shorter*
than the body, and small, beady eyes.
MEADOW VOLE   To 7½ in. long
The most widely distributed vole in North
America. Dark brown in the East to grayish
brown in the West. Found in *meadows* and
thick vegetation *near water,* from the
northern U.S. through Canada to Alaska.
YELLOWNOSE VOLE   To 6½ in. long
Grayish brown, with a distinct *yellow
patch* behind the nose. Inhabits cool,
moist, rocky woodlands from eastern Canada south to the Smokies.
PRAIRIE VOLE   To 6½ in. long
Widespread in prairies and dry areas from
the Ohio Valley to the Rockies. Common
along fencerows, railroad rights-of-way, and
old cemeteries.
PINE VOLE   To 5 in. long
Widespread in eastern and southeastern
U.S., chiefly in broadleaf forests (despite the
name). This vole is a rich *auburn* color. It
has tiny ears, and a very *short tail.*

BOREAL
REDBACK
VOLE

MEADOW VOLE

YELLOWNOSE VOLE

PRAIRIE VOLE

PINE VOLE

RODENTS: PHENACOMYS AND LEMMINGS

TREE PHENACOMYS   To 7 in. long
Bright reddish brown, with a *long, black-ish, well-haired tail.* Lives high in spruce, hemlock, and fir trees, feeding on their needles. Builds huge twig nests up to 150 ft. above ground. Found only in northwestern California and western Oregon.

MOUNTAIN PHENACOMYS   To 6 in. long
Found in grassy areas near mountain tops, rocky slopes, and coniferous forests in Canada and southwards in our western mountains. A *grayish brown* vole with white feet and a relatively short tail.

Lemmings

Five species of small, vole-like mammals, found chiefly in the tundra of the *far north.* Long, soft fur often conceals their tiny ears. Tail *very short.*

SOUTHERN BOG LEMMING   To 6 in. long
A round, brown lemming that is *almost tailless.* It makes runways just below the ground in grassy meadows of the Midwest, northeastern U.S., and eastern Canada. Feeds on grasses and clover. The **Northern Bog Lemming** (not shown) is slightly larger. It lives in bogs, mountain meadows, and tundra from northern New England west to Alaska.

BROWN LEMMING   To 6¾ in. long
Looks like an oversized vole with a grayish head and shoulders and a bright *reddish brown back and rump.* Does not turn white in winter. Ranges from Hudson's Bay west through Alaska and south to British Columbia, chiefly in tundra.

GREENLAND COLLARED   To 6⅜ in. long
LEMMING
The distinctly *patterned* summer coat of this lemming includes a *black stripe* from head to tail on the back, a pale face, and a rusty collar. *All white in winter.* Restricted to *tundra* from Hudson's Bay west to Alaska Peninsula. The similar **Hudson's Bay Collared Lemming** (not shown) lives in northern Quebec and Labrador.

TREE PHENACOMYS

MOUNTAIN PHENACOMYS

SOUTHERN BOG LEMMING

BROWN LEMMING

GREENLAND COLLARED LEMMING

RODENTS: WOODRATS

Also known as "pack rats" or "trade rats."
These rats carry small objects—including
pieces of rubbish, buttons, coins, and jack-
knives—to their nests and store them there.
Woodrats are about the same size as Old
World (non-native) rats, but have a *hairy*
(not scaly) *tail,* fine soft fur, larger ears, and
white feet and underparts. The 8 species
build a variety of huge *stick nests* in trees,
cliffs, cacti, and brush. They are generally
nocturnal and shy, but may be active and
quite bold by day in caves and abandoned
mine tunnels. Their food consists of
insects, snails, buds, leaves, fruit, and
fungi.

EASTERN WOODRAT   To 17 in. long
A large, *grayish brown* woodrat, darker
than the others. Ranges through southeastern U.S. and southern plains, living in cliffs
in the Appalachians and wooded swamps in
the South.

WHITETHROAT WOODRAT   To 15½ in. long
Found in *deserts* of the southwestern U.S.,
chiefly in Arizona, New Mexico, and west
Texas. Feeds on cactus, mesquite beans,
and various seeds. Builds nests in
extremely thorny cholla or prickly pear cactus.

DESERT WOODRAT   To 13½ in. long
Smaller and *paler* than the whitethroat.
Found in deserts and rocky slopes of Cali-
fornia, Nevada, Utah, and western Arizona.
Builds nests on ground or along cliffs.

BUSHYTAIL WOODRAT   To 17 in. long
Grayish tawny to almost black above; easily
identified by its long, *bushy, squirrel-like*
tail. Widespread in pines and rocks of western mountains from Grand Canyon and
Black Hills to Yukon.

EASTERN
WOODRAT

WHITETHROAT
WOODRAT

DESERT
WOODRAT

BUSHYTAIL
WOODRAT

OLD WORLD RODENTS

These invaders from Asia and Europe have adapted well to people's habits and habitats. They live in our buildings and ships, steal our stored food, and thrive on our garbage-rich society. Great nuisances and carriers of disease, these rodents can be told from our native mice and rats by their *long, hairless tails.*

HOUSE MOUSE   To 7 in. long

A small, *grayish brown* mouse. Lacks the contrasting white underparts of most native mice. Note its *scaly tail,* about as long as the body. This mouse lives in houses and other buildings throughout settled North America; it is occasionally found in cultivated fields.

NORWAY RAT   To 18 in. long

Also known as the Brown, Sewer, or House Rat. It originated in central Asia, spread across Europe between 1500 and 1700, and was brought to North America by ship around 1776. Grayish *brown* above, with gray (not white) underparts. Its long, scaly, naked tail is *shorter* than its body. Feeds on garbage, insects, stored grain, and will kill chickens. An excellent digger, making its own tunnels under buildings and dumps. Found throughout the lower 48 states and southern Canada.

BLACK RAT   To 18 in. long

Also known as the Ship or Roof Rat. Invaded Europe from southeast Asia centuries before the Norway Rat; brought to North America in the early 1600s. Displaced by the larger, more aggressive Norway Rat in many areas. Found chiefly in seaports, the southern U.S., and on the roofs of buildings. Two color phases—*brown* and *black.* The tail is *longer* than the body.

HOUSE MOUSE

NORWAY RAT

Norway Rat
walking

1½ in.

r.h.   r.f.

brown phase

black phase

BLACK RAT

RABBITS: HARES

Hare and Rabbit Family

In comparison with the pikas (p. 98), hares and rabbits have long ears, long hind limbs, short "cottony" tails, and bulging eyes. In this group of mammals the females are larger than males. Although hares and rabbits are usually silent, they may squeal and often thump the ground with their hind legs to communicate danger.

Hares, including the misnamed "jack-rabbits," are larger than rabbits and have longer ears. Their more powerful hind limbs enable them to leap distances of up to 20 ft. and run at speeds up to 35 mph. They make no nests, and their young can hop about within hours of birth.

WHITETAIL JACKRABBIT To 26 in. long
Lives in open, grassy (or sagebrush) plains and barren fields of the northern U.S. and the Canadian prairies, from Wisconsin and Kansas west to the Cascades and the Sierra Nevada. *Brownish gray* in summer, with a *white tail*. In winter it is often *white* or pale gray, with black-tipped ears. Chiefly nocturnal, it feeds on grasses in summer and on buds, bark, and twigs in winter.

BLACKTAIL JACKRABBIT To 25 in. long
Common in open prairies, fields, and deserts of southwestern U.S. north to Washington and South Dakota, including most of California and Texas. Grayish brown with large, black-tipped ears. The *top of the tail* and rump are *black.*

ANTELOPE JACKRABBIT To 26 in. long
Dark brown back and rump *contrasts* with dazzling white sides. The enormous 8-inch ears lack black tips. Found in saguaro forests and cactus flats of southern Arizona. Feeds on cactus, coarse grasses, and seeds from late afternoon to midmorning.

winter

WHITETAIL
JACKRABBIT

summer

BLACKTAIL
JACKRABBIT

ANTELOPE
JACKRABBIT

Jackrabbit

7–12 ft.

l.h.

l.f.

2¾ in. +

RABBITS: NORTHERN HARES

SNOWSHOE HARE   To 26 in. long
Dark brown in summer, including the tail,
becoming *white* with black-tipped ears in
winter. Also called the Varying Hare. In
spring and autumn it often appears
"patchy." It is found in northern forests,
swamps, and thickets of Alaska and Canada
southward to the Great Lakes, Rockies, and
Appalachians. Does not hibernate. Survives
the winter by feeding on buds, twigs, bark,
and frozen carrion. Its wide feet act as
"snowshoes." Known for its boom and bust
population cycles.

ARCTIC HARE   To 26 in. long
This hare lives on the arctic tundra of far
northern Canada, ranging south to Hudson's Bay and Newfoundland. *Grayish
brown* in summer, it turns *white* in winter
(except for its black-tipped ears). This hare
differs from the Snowshoe Hare by its *white
tail* in summer and its larger size. On Baffin
Island it remains white all year. Active year
'round and often found in large groups. It
often stands on its hind feet and hops with
its forefeet held against its chest. The **Tundra** or **Alaskan Hare** (not shown) is similar.
It lives in tundra and thickets of northern
and western Alaska.

EUROPEAN HARE   To 27 in. long
A non-native hare that now lives in the wild
in the northeastern U.S. and southern
Ontario. Larger than any other hare or rabbit in its range. Inhabits open fields in hilly
country. It escapes predators by making
wild, zigzagging dashes and doubling back
on its tracks. The body is brownish in summer and grayish in winter, and the *tail is
black above.*

Snowshoe Hare

r.h.

r.f.

6 in.

1–10 ft.

winter

summer

SNOWSHOE
HARE

winter

ARCTIC
HARE

summer

EUROPEAN
HARE

EASTERN RABBITS

Rabbits are smaller than hares, with shorter ears and hind legs. Although they are good runners, they usually try to hide in thickets to escape enemies. They make nests of their own fur, grasses, and leaves, where they raise their young. The young are born naked with their eyes closed and need weeks of care.

EASTERN COTTONTAIL   To 19 in. long
Grayish brown mixed with black hairs above and a *rusty nape.* Top of feet *whitish* and the cottony tail is *white.* Widespread from the Atlantic Coast west to the Rocky Mountains and Arizona, in heavy brush, mixed woods and fields, swamp edges, and weed patches. It can damage gardens, shrubs, and small trees. The **New England Cottontail** (not shown) inhabits mountains in the eastern U.S. It usually *lacks the rusty nape,* has a *black patch* between the ears, and is *redder.*

MARSH RABBIT   To 18 in. long
Restricted to southeastern bottom lands, swamps, and hummocks from coastal Virginia through Florida. It is *dark brown* with *small feet,* which are *reddish brown* above. Its tail is *small,* inconspicuous, and dingy white. Chiefly nocturnal, it searches for food such as bulbs, cane, and grasses in marshes. Escapes predators by leaping into water and swimming with only its eyes and nose above surface.

SWAMP RABBIT   To 20 in. long
Our *largest* cottontail. Replaces the Marsh Rabbit from Georgia west to Texas and north to southern Illinois. Brownish gray, *mottled* with black above, it is larger than the Marsh Rabbit and has *paler rust* fur on the tops of its feet. The Eastern Cottontail has whiter hind feet and a more distinct rusty nape. The Swamp Rabbit is found in swamps, marshes, and bottom lands and is often seen swimming in water. It forages in water and ashore on cane and crops.

EASTERN
COTTONTAIL

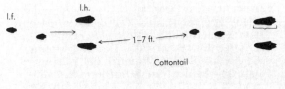

l.f.   l.h.

←→ 1–7 ft. →

Cottontail

MARSH RABBIT

SWAMP RABBIT →

WESTERN RABBITS AND PIKA

DESERT COTTONTAIL To 16 in. long
The common cottontail of low valleys, open plains, and foothills from the Western Great Plains through the arid Southwest to the Pacific in California. The body is *pale gray washed with yellow.* The ears are relatively long. This cottontail can climb sloping trees.

MOUNTAIN COTTONTAIL To 25 in. long
Paler than the Eastern Cottontail, and *grayish* in color. Its *black-tipped ears* are shorter than those of the Desert Cottontail. The *only* cottontail in much of the mountains of the West from the Sierra Nevada and New Mexico north to Alberta. Inhabits thickets, sagebrush, rocky areas, and cliffs.

BRUSH RABBIT To 14 in. long
A small, *dark brown* rabbit with relatively *small ears* and *tail.* Lives in *chaparral* and brush of California and western Oregon only.

PYGMY RABBIT To 11 in. long
The smallest North American rabbit. *Slate-gray* with *pinkish tinge.* The *tail* is *nearly hidden* and the *ears* are *small.* Digs its own burrows among clumps of tall sagebrush in cooler deserts of the Great Basin.

PIKA To 8½ in. long
Pikas are in a *separate family* from the rabbits and hares. They have *short, broad, rounded ears* and *no visible tail.* They occur in *colonies* on open rocky hillsides high in our western mountains. Their call is a *bleat* that is very hard to trace. Look for their piles of *fresh hay* drying in the sun (in winter, it will be stored under rocks). The **Collared Pika** (not shown) of Alaska, Yukon, and northern British Columbia has a pale gray collar and a white belly.

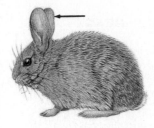

DESERT COTTONTAIL

MOUNTAIN COTTONTAIL

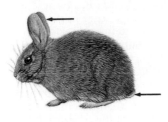

BRUSH RABBIT

PYGMY RABBIT

PIKA

HOOFED MAMMALS: DEER
Deer Family

The members of this family have 2 *toes* on each foot. The males have *antlers*, which begin to grow each spring as soft, tender bone covered with thin skin and fuzzy hair. Later in the year, after the antlers have hardened, the skin dries up and falls off. Antlers are used in courtship battles in the autumn, after which they drop off.

ELK (WAPITI)   To 5 ft. high (at shoulder)
Large deer with a *pale brown yellowish rump, short tail*, and *dark brown* legs, belly, neck, and head. Males have a *shaggy neck mane* and, in late summer and autumn, a set of large, spreading antlers. Usually seen in *groups* of 25 or more, with old males in separate groups in summer and both sexes together in winter. Feed in semi-open forests and mountain meadows in summer, descending to valleys in winter. Males bugle and battle for control of "harems" of females. The Elk has been killed off over much of its range in the East and the Great Plains and now survives chiefly in the Rockies and Northwest. Best seen in Grand Teton, Yellowstone, Olympic, Glacier, Rocky Mountain, Banff, and Jasper parks. Males weigh up to 1000 pounds.

WOODLAND CARIBOU   To 4 ft. high
Brown, shaggy fur and a *whitish neck and mane*. Males and most of the *females* grow antlers. The males' antlers are *large* and *partly flattened*. Both sexes have hooves that spread out in summer, making it easier to walk in bogs; in winter the toes close up, which gives this caribou a better grip on the ice. Although it has been killed off in much of its former range in the northern U.S., it survives in wilder Canadian forests such as Prince Albert Park, Saskatchewan. Males weigh up to 600 pounds.

BARREN GROUND CARIBOU   To 4 ft. high
Pale whitish, with a brownish wash on back. The *antlers are less flattened* than in the Woodland Caribou. Huge numbers migrate over the tundra of Alaska and the Canadian Arctic. Best seen at Denali Park in Alaska. Males weigh up to 400 pounds.

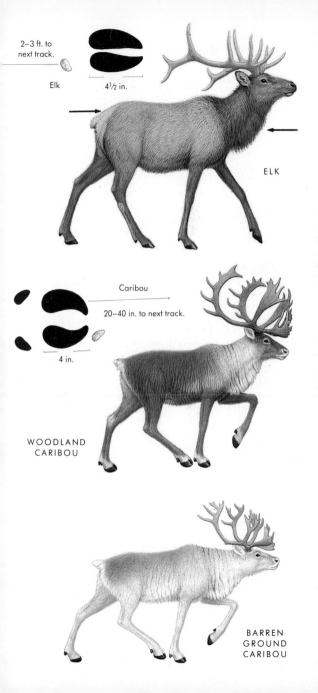

2–3 ft. to
next track.

Elk

4½ in.

ELK

Caribou

20–40 in. to next track.

4 in.

WOODLAND
CARIBOU

BARREN
GROUND
CARIBOU

HOOFED MAMMALS: DEER

WHITETAIL DEER   To 3½ ft. high (at shoulder)

Widespread in woodlands, swamps, and brush over most of the U.S. and southern Canada. A glimpse of this deer's *"white flag" tail* disappearing into a forest, or of a doe with its *white-spotted fawn* are high points of any day's outing. Although the Whitetail is *reddish* most of the year, its coat becomes *grayish* in winter. It eats twigs, grasses, fungi, and acorns. Its excellent sense of smell enables it to pick up scent of humans and move off without being seen. It is the most important big game mammal in the East, but it can become a nuisance in crop fields and orchards. Because its major predators (wolves and cougars) are extinct over much of its range, the Whitetail sometimes becomes overpopulated. Hunting can help control its numbers.

Whitetails are excellent swimmers and often are found on islands in lakes. They are also common in woody suburbs and are often struck by automobiles. Active day or night, they can run up to 35 mph, jump over obstacles 8 ft. high, and cover 30 ft. in one bound. Smaller forms of the Whitetail can be seen in the Florida Keys (**Key Deer**) and in southern Arizona.

The Whitetail's range overlaps that of the Mule Deer (p. 104) from Minnesota west. Where both occur, the Whitetail can be identified by its *smaller ears* and feet, its longer tail *(without any black above)*, a more slender neck, a narrower face, and daintier legs. The males' antlers are low and compact, with several short, *unbranched tines* on 2 main, forward-leaning beams. Males can weigh up to 400 pounds.

winter male

summer female

WHITETAIL DEER

antlers in velvet: summer
MULE DEER (see p. 104)

fawn

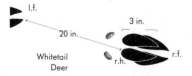

l.f.

20 in.

3 in.

Whitetail
Deer

r.h.   r.f.

HOOFED MAMMALS: DEER

MULE DEER   To 3½ ft. high (at shoulder)
Best known for its long, *mule-like ears* that
move constantly and independently as it lis-
tens for danger. The Mule Deer differs from
the Whitetail not only by its larger ears, but
also by its *shorter tail*, which is *black
above*; its thicker neck; wider face; and
stockier legs. The antlers of the male grow
higher and each branch divides into equal
forks, unlike a Whitetail's, which have
many unbranched tines. When a Mule Deer
senses danger, it leaps in an odd way called
stotting—stiff-legged, with all 4 feet off the
ground. Males weigh up to 400 pounds.

Typical Mule Deer live in the mountains,
plains, and deserts of the West, from western Texas, Arizona, and California north
through western Canada. These deer have
white rump patches, a *black-tipped tail*
above (white below), and *contrasting* blackish heads, with whiter faces and throats. In
the dense, moist forests of the Pacific Coast,
from northern California to the Alaskan
Panhandle, live 2 smaller races known as
Blacktail Deer. These deer are darker
brown. The top of the tail is *entirely blackish*, not white at the base.

The Mule Deer is the most important big
game mammal of the West. Active day or
night, it may be seen in most western parks
and protected areas. It follows definite
trails, and in mountain areas it will spend
summers in high country and retreat to valleys in winter, sometimes in groups.

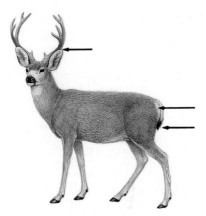

Rocky Mountains: winter

MULE DEER

Northwest Pacific Coast: winter

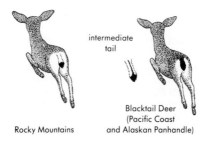

intermediate
tail

Rocky Mountains

Blacktail Deer
(Pacific Coast
and Alaskan Panhandle)

HOOFED MAMMALS:
MOOSE AND PRONGHORN

MOOSE To 7 ft. high (at shoulder)
Weighing up to 1400 pounds, the Moose is
the largest deer in the world. It is dark
brown, with *gray legs*, an *overhanging
snout*, a long *dewlap* or beard hanging
from the throat, and a *shoulder hump*.
Males grow *massive, flat antlers* with small
prongs projecting from the borders. Elk
(Wapiti) and Caribou (p. 100) have pale
rumps and lack the long face and snout.
The Moose occurs in *northern forests*,
where it is often seen feeding on water
plants far out in ponds and lakes. In winter
it feeds on twigs, bark, and saplings. It is a
fast swimmer, and on land it can run as
fast as 35 mph. The rust-colored young lack
the white spots of other deer fawns. Found
across Canada and Alaska, the northern
Rockies, upper Midwest, and northern New
England. Best seen in the following parks:
Baxter (Maine), Algonquin, Isle Royale,
Grand Teton, Yellowstone, Glacier, Banff,
Jasper, and Denali.

PRONGHORN To 3 ft. high
The only member of the Pronghorn family,
found only in the arid plains of the American West. Sometimes mistakenly called an
antelope. Best seen in Yellowstone, Wind
Cave (South Dakota), and Pawnee National
Grassland (Colorado). The Pronghorn has
true horns rather than antlers, which are
found only in the deer family. True horns
are not normally shed each year, but the
Pronghorn's horns are covered by horny
sheaths that are *shed each year*. All males
and most females have horns. Both sexes
have conspicuous *white rumps*, white
throat stripes, and white sides. Males alone
have black patches from nose to eyes and on
the neck. The Pronghorn browses on weeds,
sagebrush, and grasses. When alarmed it
raises its white rump hairs and dashes off
at speeds of over 50 mph. It is the fastest
mammal in the Western Hemisphere and
one of the fastest in the world.

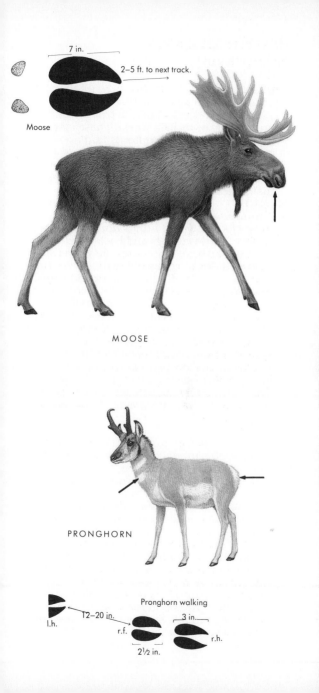

7 in.

2–5 ft. to next track.

Moose

MOOSE

PRONGHORN

Pronghorn walking

l.h.

12–20 in.

r.f.

3 in.

2½ in.

r.h.

HOOFED MAMMALS:
MUSKOX AND BISON
Wild Goats, Sheep,
and Relatives (Bovids)

This family includes the Bison, Muskox, goats, sheep, cattle, and the true antelope and buffalo of the Old World. These mammals have *unbranched horns* that are *never shed.* The horns are made up of a sheath covering a bony outgrowth of the skull. Horns are present in *both sexes,* not just males.

MUSKOX   To 5 ft. high (at shoulder)
A brown ox of the far northern tundra, with *long, silky* hair that *hangs down* nearly to its feet. The lower legs and back ("saddle") are silvery white. The broad, flat horns are plastered close to the skull, with curved tips that point forward. The female's horns are more slender and curved. Small herds graze in valleys in summer and seek windswept slopes and hilltops with less snow in winter. They usually face danger by forming a circle with their heads facing outward against wolves or human hunters. Once important as a food item for Eskimos, Muskoxen were exterminated over much of their range. They are being reintroduced in parts of Alaska and northern Canada. A Muskox can weigh up to 900 pounds.

BISON   To 6 ft. high
This enormous mammal (which can weigh up to 2000 pounds) is dark brown, with a *high hump* on its shoulders that is usually pale golden brown. The *head* is *massive* and both sexes have *horns. Long, shaggy hair* hangs from the shoulders and front legs. Before North America was settled by Europeans, Bison numbered perhaps 60 million, but by 1900 fewer than 1000 remained. A last-minute rescue attempt resulted in a successful comeback. Now you can see Bison at Yellowstone, Wind Cave (South Dakota), Wood Buffalo (Alberta), and many other parks and refuges. Bison are also called Buffalo, but true buffalo live only in Africa and Asia.

r.f.   r.h.

MUSKOX

BISON

Bison

5 in.   3 ft. or less to next track.

HOOFED MAMMALS: GOATS AND SHEEP

MOUNTAIN GOAT   To 3½ ft. high (at shoulder)
A yellowish white mammal with a mane on its throat that looks like a beard. Its short, smooth, *thin, black horns* curve slightly backward. In summer its hair becomes shorter. At that time of year it is found on rocky crags near snowline. In winter it descends to lower elevations and develops a longer, shaggier coat. An excellent climber, it has flexible black hooves, a compact and muscular body, and short legs that are ideal for balance but poor for running. The Mountain Goat is famous for making its way along incredibly narrow ledges. Small herds inhabit mountains from Montana to southern Alaska.

BIGHORN SHEEP   To 3½ ft. high
A *thick-necked* sheep with a *creamy white rump.* The Bighorn is larger and dark brown in the northern U.S. and Canadian Rockies and smaller and pale tan in south-western deserts. Males feature *massive coiled horns* that spiral back, out, and then forward, forming a full *curl.* Females have shorter, thinner horns, resembling those of Mountain Goats. Bighorns live on mountain slopes, in meadows, and in rugged rocky areas with few trees. Males engage in serious butting contests in fall, when the sounds of their crashing heads can be heard over a mile away. Bighorns are best seen at Death Valley, Yellowstone, and Glacier.

WHITE SHEEP   To 3½ ft. high
This wild sheep replaces the Bighorn in Alaska and Yukon. Often known as Dall's Sheep, it is *all white* with black hooves and massive yellowish horns that are somewhat *more slender* than those of Bighorns. It is common at Denali Park in Alaska. In southeastern Yukon and northern British Columbia a dark form known as **Stone's Sheep** occurs. This dark form varies from black to brown or silver, with a *contrasting* white rump, belly, face, and *leg trim.*

110

Mountain Goat

About 15 in. to next track.

3 in.

MOUNTAIN GOAT

3 in.

About 15 in. to next track.

Bighorn Sheep

BIGHORN SHEEP

black phase

white phase

WHITE SHEEP

HOOFED MAMMALS

PECCARY (JAVELINA)   To 36 in. long

This *pig-like* mammal belongs to the fourth family of hoofed mammals in North America. Unlike the Wild Boar, which was introduced from Europe, this wild pig is native to North and Latin America. The Peccary lives in bands of up to 25 in semi-arid deserts and hills covered with oaks, chaparral, and mesquite in southern Arizona and Texas. It feeds on prickly pear cactus (spines and all), mesquite beans, tubers, leafy plants, toads, lizards, baby birds, and even snakes.

The **Wild Boar** (not shown) has been introduced as a game mammal in California and much of the South. Its tusks are *curved* and curl *out and up,* while the Peccary's tusks are *straight* and *point downward.* The Wild Boar has a long, straight tail.

SIRENIANS

MANATEE   To 13 ft. long

The only sirenian (p. 9) in North America. A grayish, *nearly hairless* aquatic mammal with flippers and a broad, *paddle-shaped tail.* Its broad head, with split upper lips and numerous stiff bristles, is all that is usually seen above water. Manatees often hiss when they meet and embrace each other with their flippers. These mammals live in warm rivers and along the Atlantic and Gulf coasts, from North Carolina to Texas. Most are found in Florida. Almost all manatees are seen with scars from motorboat propellers.

EDENTATES

ARMADILLO   To 33 in. long

The only edentate (p. 9) in North America. It is covered with *heavy, bony armor* over the head, body, and tail. Flexible bands across the middle of its body allow it to twist and turn. Strong claws enable it to dig burrows in sandy soils or streamsides. The Armadillo feeds chiefly on insects, crayfish, frogs, bird eggs, and berries. Its range has *expanded* recently, from Texas east to Florida and north to Missouri.

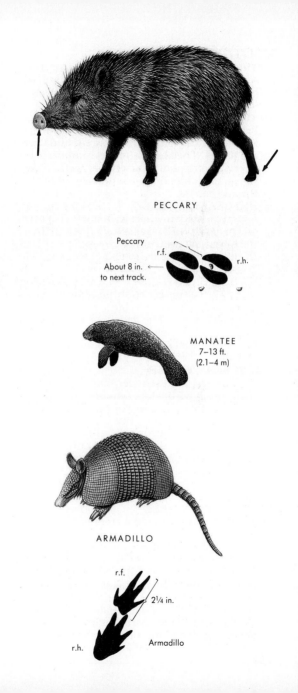

PECCARY

Peccary

r.f.

About 8 in.
to next track.

r.h.

MANATEE
7–13 ft.
(2.1–4 m)

ARMADILLO

r.f.

2¼ in.

r.h.   Armadillo

TOOTHED WHALES

This group is made up of beaked and sperm whales and the dolphins and porpoises. All have *simple, peglike teeth* and no baleen. (Compare with baleen whales, p. 122.)

Beaked Whale Family

Primitive whales, rarely seen and not well known. These whales hunt squid and deep-sea fishes. They have a *narrow snout*, usually with *only 2 or 4* functional teeth. Note the small dorsal fin *near the rear* and the 2 throat grooves.

BAIRD BEAKED WHALE   To 42 ft. long
A large, blackish whale with a whitish area on the lower belly. Has a *long beak*, with the lower jaw protruding beyond the upper. Occurs in schools of up to 30 off the Pacific Coast.

TRUE BEAKED WHALE   To 17 ft. long
One of 7 closely related, little-known whale species. This species is *slate black* above and paler below. The male has *2 small teeth* at the *tip* of the lower jaw. It is found in the North Atlantic, sometimes as far south as Florida.

GOOSEBEAK WHALE   To 28 ft. long
Also known as the Cuvier Whale. Varies in color, often with white patches on the head, back, or belly and a background color of black, brown, or gray. Has a thick body with a *distinct keel* running down the back from the dorsal fin to the tail. The snout is *roundish*, unlike the longer snouts of other beaked whales. This whale lives off both coasts. It swims in groups of up to 40. Males have 2 teeth in the lower jaw.

BOTTLENOSE WHALE   To 30 ft. long
Found only in the cold waters of the Arctic and North Atlantic oceans. Males have a *high forehead* that rises above a short beak and 2 small teeth at the tip of the lower jaw. Older males have a white dorsal fin and a white patch on the forehead. Body color varies from grayish black to light brown, with a whitish belly. The Pilot Whale (p. 120) is similar, but *black overall*, with a *larger dorsal fin* on the *front* half of the body.

114

BAIRD BEAKED WHALE
35–42 ft.
(10.7–12.8 m)

TRUE BEAKED WHALE
15–17 ft.
(4.5–5.2 m)

GOOSEBEAK WHALE
18–28 ft.
(5.5–8.5 m)

BOTTLENOSE WHALE
20–30 ft.
(6.1–9.1 m)

TOOTHED WHALES

SPERM WHALE To 60 ft. long

This whale, featured in the book *Moby Dick*, is the only species in its family. It has an enormously high, *squared-off forehead*. Its head is one-third of its length. There are up to 28 strong teeth on each side of the *long, narrow lower jaw*. The back has *no dorsal fin*. When this whale exhales or "blows" at the surface, its spout of mist is *directed forward*. Able to dive down to 2 miles below the surface, the Sperm Whale feeds on giant squid, octopus, bottom-dwelling sharks and other fish. Found off both coasts, it has been overhunted for its oil, which is used to lubricate machines.

PYGMY SPERM WHALE To 13 ft. long

This whale and the related **Dwarf Sperm Whale** (not shown) are the only small whales with a *forehead (snout) that protrudes* in front of the mouth. The Pygmy is black above the grayish white below, with a pale, *bracket-shaped mark* behind each eye. It has a small dorsal fin. The lower jaw is narrow, with many needle-like teeth. Found in warm waters off both coasts.

WHITE WHALE (BELUGA) To 14 ft. long

This is a small *white* whale of cold arctic waters that feeds on fish and cuttlefish. Note its fairly *high* forehead, *very short snout*, and *lack of a dorsal fin*. It feeds in shallow waters and rivers from Alaska to Hudson's Bay and south to the St. Lawrence River. Young are born brown and gradually turn gray, then white by their sixth year.

NARWHAL To 15 ft. long

Both sexes are *mottled brownish above* and paler below. They have *no snout* and *no dorsal fin*. Narwhals live in the high Arctic, in cold seas from northern Alaska to Greenland. They feed on fish and deepwater squid and crustaceans. Both sexes have only 2 upper front teeth. In males the left one grows into a long, hollow, twisted, *unicorn-like tusk* that can reach 9 ft. long.

SPERM WHALE
40–60 ft.
(12.2–18.3 m)

PYGMY SPERM WHALE
9–13 ft.
(2.7–4 m)

WHITE WHALE
11–14 ft.
(3.4–4.3 m)

NARWHAL
12 ft.
(3.6 m)

WHALES: DOLPHINS

These small whales usually have a *well-developed dorsal fin* and notched tail flukes. They commonly travel in large groups and often ride the bow waves of ships.

SPOTTED DOLPHIN   To 7 ft. long

Fairly common in the Gulf of Mexico and northward to North Carolina. Numerous *large white spots* on a blackish back. The long snout is separated from the forehead by a distinct groove. The **Striped Dolphin** (not shown) is black above and white below, with *black stripes* from eye to flipper and undersides. It is found on both coasts in colder waters.

COMMON DOLPHIN   To 8½ ft. long

Found off both coasts, this dolphin is black above with black flippers, contrasting with its *yellow flanks* and white belly. Note the *long beak* and black "spectacles."

PACIFIC BOTTLENOSE DOLPHIN   To 12 ft. long

Large and uniformly *grayish,* with a paler belly and a short beak with a slightly longer lower jaw. This Californian species has *white* on the *upper lip.* The **Atlantic Bottlenose Dolphin** (not shown), *common* along the Atlantic Coast, is similar but has *no white* on the upper lip.

RIGHT WHALE DOLPHIN   To 8 ft. long

Small and black, with a distinct *white belly stripe* from breast to tail and *no dorsal fin.* Lives in the Pacific from the Bering Sea south.

PACIFIC WHITE-SIDED DOLPHIN   To 9 ft. long

Greenish black above, with *whitish sides* and belly. *Blunt-nosed*—lacks the pointed beak and pale face of the Common Dolphin. Found all along the Pacific Coast. The similar **Atlantic White-sided Dolphin** (not shown) is found on the Atlantic Coast south to Cape Cod.

WHITEBEAK DOLPHIN   To 10 ft. long

A black dolphin with a *pale flank patch* and a *short white beak.* North Atlantic west to Labrador.

SPOTTED DOLPHIN
5–7 ft.
(1.5–2.1 m)

COMMON DOLPHIN
6½–8½ ft.
(1.9–2.6 m)

PACIFIC BOTTLENOSE DOLPHIN
10–12 ft.
(3–3.6 m)

RIGHT WHALE DOLPHIN
5–8 ft.
(1.5–2.4 m)

PACIFIC WHITE-SIDED DOLPHIN
7–9 ft.
(2.1–2.7 m)

WHITEBEAK DOLPHIN
7–10 ft.
(2.1–3 m)

WHALES: DOLPHINS

KILLER WHALE (ORCA) To 30 ft. long

This huge dolphin is black above and *white below*, with white extending up the sides. Note the *white oval* behind each eye, the *long dorsal fin*, blunt nose, and long flippers. Found on both coasts. It feeds on large fish, but specializes in attacking seals, sea lions, and other whales.

GRAMPUS To 13 ft. long

A large dolphin with a bulging, *yellowish head*, no beak, and mottled gray, *slender flippers*. Males are blue-gray above, females brownish. This dolphin has a much larger dorsal fin than the Goosebeak Whale. It is found off both coasts.

FALSE KILLER To 18 ft. long

A large, *all-black* dolphin. Note its *rounded head*, *large teeth* on both jaws, and relatively *small dorsal fin*. Travels in very large schools that sometimes strand themselves on beaches. Found on both coasts, from North Carolina and Washington south.

COMMON PILOT WHALE To 28 ft. long
("BLACKFISH")

A huge, *all-black* dolphin with a *bulging forehead* and a *large, swept-back dorsal fin*. It travels in large schools in the Atlantic as far south as Virginia.

HARBOR PORPOISE To 6 ft. long

A small dolphin with a *triangular dorsal fin*. It is found all along the Pacific Coast and on the Atlantic Coast south to New Jersey. Common close to shore and in harbors. Note its round mouth, black back, *pink sides*, white belly, and the *dark line* between mouth and flipper.

DALL PORPOISE To 6 ft. long

All-black, with a striking *white area* on the sides and belly. Note its blunt head and triangular dorsal fin. Found on the Pacific Coast from Alaska south to California.

120

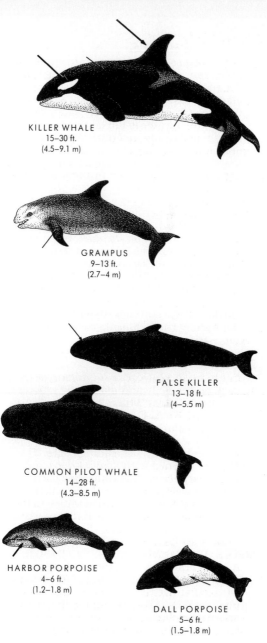

KILLER WHALE
15–30 ft.
(4.5–9.1 m)

GRAMPUS
9–13 ft.
(2.7–4 m)

FALSE KILLER
13–18 ft.
(4–5.5 m)

COMMON PILOT WHALE
14–28 ft.
(4.3–8.5 m)

HARBOR PORPOISE
4–6 ft.
(1.2–1.8 m)

DALL PORPOISE
5–6 ft.
(1.5–1.8 m)

BALEEN WHALES

Large whales with *no teeth* but with *strips of baleen*, or "whalebone," hanging from the edges of the upper jaw. These comblike strips are fringed on the inside and are used to strain food from sea water. When a mouthful of water is forced through the baleen, zooplankton and other small animals are trapped on the fringes and then swallowed. Baleen whales have 2 blowholes. Some species have been hunted to near-extinction.

GRAY WHALE   To 45 ft. long

The only member of its family. A medium-sized, blotched, grayish black whale. It has a series of *bumps on its back*, but *no dorsal fin*. Unlike finbacks (below), it feeds close to shore among rocks and kelp. The Gray Whale raises its young in shallow lagoons off Baja California and western Mexico and migrates north to the Arctic Ocean near Alaska for the summer. Its annual migrations along the California coast attract thousands of whalewatchers.

Finback Whale Family

These whales have a *small dorsal fin* far back on the body and many long grooves on the throat. (Gray Whales have only 2−4.)

HUMPBACK WHALE   To 50 ft. long

A chunky black whale with a white underside. It has *very long, mostly white flippers* with knobs along the leading edges. The head is adorned with *fleshy knobs* and white barnacles. This whale has been severely overhunted, but it can still be seen off both coasts. Lucky whalewatchers have seen the spectacular leaps it makes when breaching and slamming the surface. Known for its lovely, haunting songs.

MINKE (PIKED) WHALE   To 30 ft. long

A small finback whale, found along both coasts. Blue-gray above and white below, with a distinctive *white patch* on its flippers. Its dorsal fin has a *curved tip*. Often approaches boats, sometimes closely enough that its *white baleen* may be seen.

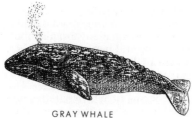

GRAY WHALE
35–45 ft.
(10.7–13.7 m)

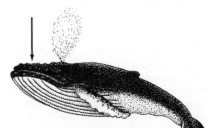

HUMPBACK WHALE
40–50 ft.
(12.2–15.2 m)

MINKE WHALE
20–30 ft.
(6.1–9.1 m)

BALEEN WHALES

SEI WHALE (RORQUAL) To 50 ft. long
This flat-headed baleen whale is larger than
a Minke and smaller than a Finback. It is
dark above and has a relatively *large dorsal
fin.* Its tail flukes are *all dark,* not white on
the underside, as in a Finback. Its *baleen* is
black with white, frayed edges. Unlike the
Minke, it has no white patches on the upper
side of its flippers. The Sei is found off both
coasts. It migrates south to warmer waters
for the winter.

FINBACK (FIN) WHALE To 75 ft. long
A very large, flat-headed baleen whale. It is
paler gray above than the Sei and the
undersides of its flippers and tail flukes are
white. The baleen is *streaked purple and
white* and the mouth is unique—the right
side of the lower lip is white and the left is
gray. The Finback is an extremely fast
swimmer. Unlike the Humpback, this whale
only rarely leaps clear of the water. It is
most common in subarctic waters but
ranges southward along both coasts, usually well offshore and often in groups.

BLUE WHALE To 100 ft. long
Weighing up to 200 tons, this whale is the
largest animal to have ever lived, dwarfing
even the largest dinosaurs. It is *blue-gray*
above, with *yellowish white* on the belly. It
has a small dorsal fin, white on the undersides of the flippers, and *black baleen.*
1500 Blue Whales summer in the Gulf of
Alaska and the Aleutians, migrating south
to waters off western Mexico for the winter.
These behemoths are seriously endangered.
The few hundred survivors off our East
Coast summer from the Gulf of St. Lawrence north along Davis Strait, wintering at
unknown places in the tropical Atlantic.
Blue Whales feed chiefly on shrimp-like
krill, strained from sea water by their comblike baleen.

124

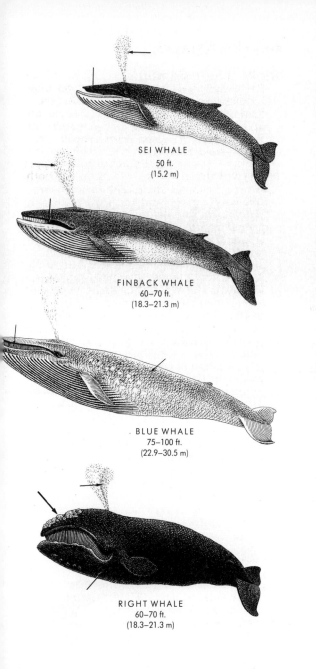

SEI WHALE
50 ft.
(15.2 m)

FINBACK WHALE
60–70 ft.
(18.3–21.3 m)

BLUE WHALE
75–100 ft.
(22.9–30.5 m)

RIGHT WHALE
60–70 ft.
(18.3–21.3 m)

BALEEN WHALES

Right Whale Family

Early whalers named these the "right" ones to kill because their oil-rich bodies would float after they were harpooned. They have an *enormous head.* The arched mouth has *several rows of baleen,* with a gap at the snout. *No dorsal fin.*

RIGHT WHALE To 60 ft. long

A large, blackish whale, with raised, *wart-like knobs* on its head and sometimes with *pale patches* on its underparts. Only a thousand individuals survive off both coasts. The Atlantic population migrates north to the Gulf of St. Lawrence for the summer, the Pacific whales to southern Alaska. Both groups migrate to warmer waters for the winter. The Right Whale's *spout* resembles a V, forming 2 columns up to 15 ft. high.

BOWHEAD WHALE To 60 ft. long

Replaces the Right Whale in cold arctic seas from western Alaska through northern Canada to the Gulf of St. Lawrence. Usually found near the edge of a polar ice field. The *huge head* has a *bowed lower jaw,* which is *mostly white.* Bowheads remain in arctic waters year 'round and are able to break through ice several feet thick.

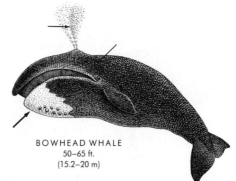

BOWHEAD WHALE
50–65 ft.
(15.2–20 m)

INDEX

The illustrations generally appear on the page facing the text.

Praise for
THE PRICE

"It scared, exhilarated, uplifted, frenzied, and made me green
with jealousy." —Ken Bruen

"Sokoloff is simply amazing." —Bookreporter.com

"[A] sublime second novel...Rest assured that Sokoloff will suffer none of the signs or symptoms of a sophomore slump with
this confident follow-up to her Stoker-nominated debut...Her
gooseflesh-inducing imagery jumps right off the pages, and her
rich, graceful prose calls to mind names like King, Saul, and
Levin." —*Dark Scribe Magazine*

"Sokoloff is on target with her portrayal of terror in the loss of a
loved one and her atmosphere of suspense, including the
lengths to which people will go to save their family members."
—*VOYA*

"Sokoloff's straightforward writing style perfectly enhances her
chilling and mysterious novel, in which she blurs the lines between what is real and what is merely a hallucination."
—*Romantic Times BOOKreviews*

"This suspenseful novel reads much like an episode from *The
Twilight Zone*, a real situation in a surreal environment. *The
Price* is fascinating, absorbing, and hard to put down."
—*Hidden Staircase Mystery Books*

"[A] fast-paced thriller in which Sokoloff executes her plot with
razor-sharp timing and skill. [A] gripping read full of questions
about good, evil, and human nature." —*Rue Morgue* magazine

MORE...

"More than an impressive debut novel; it is an impressive work, period . . . Highly recommended."   —BookReporter.com

"A top-notch gripping ghost story from the first page."
                                                    —*New Mystery Reader*

"Sokoloff's screenplay training pays off with this economical and very scary tale of five alienated teens who get caught up in forces far, far beyond their control, but she scores extra points for the Biblical twist that really ratchets up the tension."
                                                    —Sarah Weinman

"You won't be able to put this gripping page-turner down."
                                                    —*BookLoons*

"A wild and frightening journey that explores an evil that most of us might not know about."   —*Full Effect Magazine*

"An eerie ghost story that captivates readers from page one. The author creates an element of suspense that builds until the chillingly believable conclusion. Four stars."
                                                    —*Romantic Times BOOKreviews*

"Scary without being gory, this book has just the right blend of psychological drama, mystery, romance, and creepiness."
                                                    —*VOYA*

"*The Breakfast Club* meets *The Shining* in this engrossing, hard-to-put-down novel."   —Armchairinterviews.com

St. Martin's Paperbacks Titles
by Alexandra Sokoloff

The Harrowing

The Price

THE PRICE

ALEXANDRA SOKOLOFF

St. Martin's Paperbacks

This is a work of fiction. All of the characters, organizations and events portrayed in this novel are either products of the author's imagination or are used fictitiously.

THE PRICE

Copyright © 2008 by Alexandra Sokoloff.
Excerpt from *The Unseen* copyright © 2008 by Alexandra Sokoloff.

Cover photograph of hospital © Kevin Muggleton / Corbis
Cover photograph of tree © David Roth / Getty Images

For information address St. Martin's Press, 175 Fifth Avenue, New York, NY 10010.

Library of Congress Catalog Card Number: 2007040917

ISBN: 0-312-35750-8
EAN: 978-0-312-35750-4

Printed in the United States of America

St. Martin's Press hardcover edition / February 2008
St. Martin's Paperbacks edition / December 2008

St. Martin's Paperbacks are published by St. Martin's Press, 175 Fifth Avenue, New York, NY 10010.

10 9 8 7 6 5 4 3 2 1

For my family: Alexander, Barbara, Elaine, and Michael…
…and Michael

THE PRICE

PROLOGUE

Dead of winter, and snow falls like stars from a black dome of sky. All sound is swallowed by the swirling white chaos.

No human life out there on this night. The city of Boston sleeps in the storm. . . .

But underneath the ice beats a great heart that is never still.

Beneath the falling snow, a vast complex sprawls like a frozen spider buried in the white drifts—the architectural wonder of Briarwood Medical Center: six state-of-the-art hospitals symbiotically entwined. Labyrinthine underground tunnels and high glass bridges above the snow-swept streets mate the white marble, Doric buildings of the old Massachusetts Bay Medical College, the dark brick buttresses of Mercy, the sleek curves of Briarwood Children's Medical. Torturously twisting corridors wind through Gothic arches and classic Colonials and angular modern structures, creating a bewildering, futuristic maze.

Inside, the hospital has a peculiar vacuum quality of silence. In the fluorescent halls, medical personnel walk in measured paces; dazed, dreamy patients in robes drift past the open doors of darkened rooms. Snow flies outside the windows, beating soundlessly against the glass.

Deep within the labyrinth, a man moves in the endless halls: tall and dark, a graceful shadow against the white of the walls.

He is at home here—his movements fluid and unhurried, his angular face thoughtful and intent.

The corridors twist and turn, drawing the man deeper into the hospital, past vast wards with the injured and terminally ill moored in their beds. There is a throbbing pulse around the man, the heartbeat of the hospital: life-support machines augmenting labored breathing, soft moans of pain, quiet sobbing . . . and a whispering, barely audible at first, but increasing . . .

The man cocks his head slightly, listening.

The sound builds around him . . . the prayers of relatives keeping vigil . . . pleas in all languages . . . overlapping . . . rising and falling in waves . . . through anger, through tears:

Please, God . . . please help her . . . Don't let him die . . . Dear Lord . . . Signora, aiutami . . . Hear me, Jesus . . .

The dark man closes his eyes, listening to the music of the voices. Then his face sharpens, eyes opening and focusing to pinpoints, at the sound of one fierce, stark vow:

I would do anything.

CHAPTER ONE

Deep in the heart of Briarwood Children's Medical—or perhaps it was Carver Women's, the boundaries between Briarwood's separate hospitals having so merged by now, it was at times
impossible to tell the difference—stretched a long corridor rarely
traveled in the winter months.

A glass wall ran along one whole side of the corridor, looking
out on the hospital garden, in mid-March still an arctic wasteland,
the shapes of statues and trees frozen and drifted in snow. On the
other side of the hall, arched wooden doors led to the hospital
chapel. The doors were not immediately apparent or even easy to
access, but not a few people found themselves there almost by
magic, in the course of desperate midnight wanderings through
the hospital maze.

Inside, the chapel was small and dim, with four rows of
wooden pews and a low platform serving as a dais, and cold—as
if the oppressive overheating of the hospital had not been able to
penetrate here. In a center pew, Will Sullivan sat alone in the enveloping silence. Handsome in the most well-bred of ways, a classic, uniquely American combination of movie-star elegance and

frontier ruggedness, he currently looked ten years older than his forty-two years. His six-foot-plus frame seemed as stooped as an old man's, his gray-blue eyes sunken, his face haggard with worry.

Will clasped his too-dry hands as if in prayer, tried to sit up straight, but it was a great effort; he felt scraped raw, nearly dead with exhaustion. In fact, for days, or weeks, or even months, he had not been entirely sure if he was awake or asleep.

Behind the podium at the front of the chapel, a tall stained-glass window portrayed a slightly cubistic Christ as the Shepherd, watching over lambs. Against another wall, a wooden wheel depicting symbols from the world religions was mounted above a bookshelf lined with religious texts in various languages. Votive candles in red glass flickered on a side altar.

Will gazed up through bleary eyes at the patterned glass before him. Black words were scattered almost randomly in the panels, like code, and for a moment Will lost himself, puzzling over the sentence.

THE   LORD
IS   THE
STRENGTH   OF MY
LIFE   A VERY   PRESENT
HELP   IN   TROUBLE
OF   WHOM   SHALL   I
BE   AFRAID
?

Will stared harder, caught by the final phrase, the last words set apart from the others, black and grim.

BE   AFRAID

He shivered in the unheated chapel.

A shadow moved to the side of him. Will twisted in his seat, startled.

A round-cheeked, salt-and-pepper-haired chaplain stood in a side doorway, looking at Will inquiringly.

"May I help you?"

Will briefly took in the chaplain's ruddy, eager face, the wrinkled suit, the too-tight collar around the clergyman's substantial neck. Without thinking, Will shook his head. "No. Thank you."

The chaplain hesitated, but when Will turned back to face the dais, the clergyman disappeared back through the side door.

Will sat again in the silence and spoke aloud, surprising himself.

"God."

He stopped, confused.

God who?

His tired mind paged through memories of Sunday services: sumptuous cathedrals with well-heeled parishioners; midnight Masses at lace-curtain Irish churches; wakes, baptisms, charity events . . . all such a pillar of his father's political life.

There are no atheists in foxholes. Or on the campaign trail, either.

But faith? Actual faith?

Had he ever believed, Will's father?

Had Will?

Had it ever even occurred to Will that he *didn't* believe?

His mind reached for God and found—nothing. He believed in goodness, and morality, and law, and love—oh yes, love at first sight, romantic love, love of state, of country, love of justice. But God? Only in the most abstract of ways, and perhaps not even that.

Yet he had unquestioningly continued the family tradition, the Sunday-service photo op. And dragged his wife and daughter into it, Joanna never protesting, everything always for him.

Had there ever been any God under the politics?

And now, when actual miracles were required . . .

He felt on the verge of drowning. And wouldn't it be a relief to give in, to let his mind go and slip into an ocean of oblivion, unconsciousness, insanity . . .

A sudden, live stirring inside his suit jacket roused him back to the present. A small, furry nose poked out of his lapel, followed by huge dark eyes, long white ears. A rabbit.

Will felt its tiny heart racing against his own. He stroked it absently, looked up at the stained glass again, and a jolt of adrenaline spiked through him, the awareness of why he was there returning. He swallowed through a dry mouth, tried once again to pray.

"I can't . . . I can't lose her."

He could feel his heart beat, slow spasms in his chest. Only silence answered him.

After an endless, empty moment, Will rose with effort, turned away from Christ's frozen image in glass . . .

. . . and was startled to see he was not alone. Another man sat a few pews back, older than Will, yet somehow ageless, with deep-set eyes and dark hair. He must have been startlingly beautiful as a young man—a face Roman in its nobility, the chiseled-marble features powerfully masculine, but with an almost feminine sensuality of mouth; blue-black hair glinting with silver, slate-gray eyes, long limbs, and tapered fingers—his Hermès suit, his bearing, all understated, faintly European elegance.

The man's gaze lowered to Will's chest, and he smiled slightly. Will remembered the rabbit, realized how strange a picture he must present. He tucked the bunny gently back into the carrying bag inside his lapel.

"My daughter"—his voice caught on the word, and he had to swallow—"loves rabbits."

The man nodded gravely, without surprise.

"It's Will Sullivan, isn't it." It was not a question, and for a moment Will tensed, warning bells going off.

Reporter.

Just as quickly, he dismissed the thought. The man had none of the scruffiness of a journalist or the camera-ready vacuousness of a television reporter.

The man continued, gently. "I'm sorry to see you here." Will met the dark man's eyes and saw his own pain reflected there, before the other man drew a breath and his gaze became neutral, formal again, like a veil of gauze drawn over a wound.

"Your father was the best governor this state ever had. I expect you'll be better."

Will felt the heat of recognition in his chest at the man's words. Hadn't his whole life been guided by exactly that conviction? *I can be better than my father.*

Automatically his eyes warmed, his campaign smile lit his face. "I appreciate that."

The man's gaze was steady, and Will thought: *He knows what bullshit that is. It's all so irrelevant now. . . .*

The man glanced up at the stained glass of the Christ; the look on his face was ambiguous, rueful.

"There is a way," he said, his voice low—so low that Will frowned, not sure if he'd heard right.

"I'm sorry?"

But the man merely nodded courteously, almost a bow. "I wish you . . . the best." He withdrew discreetly, moving out of the chapel with a whisper of doors. Will noted the heaviness about him, the effort with which he moved despite the elegant carriage, and wondered why he had ever thought the man was anything but what Will himself was: a desperate relative, come to bargain with a mythical God for a miracle.

He turned back to look at the glassy Christ. His body sagged, his head dropped to his chest, as he whispered hoarsely:

"Please."

CHAPTER TWO

It had been like fate, a fairy-tale curse, mythic in its construction. An impossibly beautiful day, glorious, the air crisp with fall and brilliant with sun; trees flaming with color in the Common; dogs and seagulls and squirrels sharing the paths with in-line skaters, lovers, parents with strollers; the whole garden teeming with life.

The crowd that gathered that day in a circular green was hundreds more than Will's campaign staff had even dared to hope. There was a happy, family, American aura to the event: an outdoor band shell with a small stage festooned with red, white, and blue bunting; sweet-faced senior citizens standing behind long tables serving apple cider and Krispy Kreme, clowns handing out helium balloons to jostling children—all a bit of old-time, small-town U.S. of A., framed by a modern city skyline.

Brilliant camera flashes rippled through the crowd; reporters jostled and commented from the sidelines while Senator Flynn, Irish working-class hero, American political institution, longtime blood brother of Will's father, gave Will a glowing introduction: District Attorney Sullivan, ten years as a prosecutor, four on the city council, fighter for right, defender of the weak, prince of the blood, soon to be king.

By Will's side, always, was the most beautiful woman in the world—darkly lovely, deeply mysterious: his wife, Joanna; and between them, their daughter, Sydney, a sparkling, imperious five-year-old, radiant and basking in the attention, yellow balloon bobbing above her from a ribbon tied around her tiny wrist.

And hovering in the wings, the kingmakers: politicos who had held court in the shadow of Will's father, a powerful if uneasy alliance of Irish political aristocracy and blue-blood patrons from his Brahmin mother's circles, watching Will now with a predatory intensity. Will had known all too well what they were whispering:

The very definition of shoo-in . . . the man can go all the way. . . .

And while he knew he stood there partly because of his name and pedigree, he also knew his name was the gift that would allow him to begin clean and stay clean—to do some good in the world without having to sell his soul for the chance.

Then the applause rose as the senator's voice boomed through the park, and Will jogged out onto the platform—and onto the national stage: "Ladies and gentlemen, I give you the next governor of Massachusetts—Will Sullivan."

At home that night, the applause still rang in Will's ears as the sun went down in fiery glory outside their Tudor mansion in the woods.

He remembered Joanna taking Sydney upstairs, and the backward look she gave him: heart-stopping, full of promise.

And the phone had rung—Jerry, his campaign manager, rhapsodizing about national news coverage and polling points.

And then the moment Will had visualized over and over, that haunted his dreams: Sydney and Joanna singing together in the steamy bathroom, where Joanna bathed Sydney in the claw-foot tub, their faces shining, dewy with sweat. And then Sydney flinching in pain, pulling away from the sponge.

Joanna's surprise, her frown, as her fingers moved tentatively over her daughter's stomach . . . the stab of fear as she found the hard, alien mass . . .

While downstairs, Will listened to Jerry, knowing that nothing was certain, that the race had just begun, but for a moment allowing himself the dream . . .

. . . and Jerry's words on the phone: "Nothing can stop us now. . . ."

Then Joanna standing in the doorway, holding Sydney, dripping, in a towel.

And Will dropping the phone.

And seeing that his life as he knew it was over, as he looked into the terror in Joanna's eyes.

CHAPTER THREE

The elevator jolted to a stop. Will jerked upright, opened his eyes, wondered briefly if he'd actually fallen asleep standing up.

The doors slid open, and he walked out into the Briarwood Children's Medical cancer ward.

It had somehow turned into morning. Bleak winter light through the wide windows of the playroom illuminated a scene from Grimm: pale, skeletal children drifting under towering papier-mâché sculptures—a grazing giraffe, a bulbous frog, a stalking tiger. Huge butterflies hung from the ceiling; oversize alphabet blocks were tumbled in a corner; there was a dreamlike unreality about the scene, the children like gnomes, bald and gaunt, propelling themselves in wheelchairs, pushing around IV carts with tubes snaking into their arms.

As Will walked through the wide round room, he straightened, shrugging off his own distraction to nod and smile at other parents. They knew one another by sight, sometimes by name, but generally kept their distance, walled in by their own grief.

He stopped beside a young, fragile mother leaning against the wall, biting ragged nails to the quick as she watched several children coloring with chalk at a low table, her eyes clinging to a tiny, bald boy.

"How's Eli?"

She jumped, but smiled wanly as she recognized Will, answered bravely, "We're having a good day today."

"I'm so glad."

Will heard his father's enveloping charm in his voice and hated himself for how easily he slipped into politician mode. But the young mother's eyes brightened with gratitude as he touched her shoulder briefly in support, and she looked a little lighter as he turned away to leave. So it wasn't all a game, not entirely. It didn't matter that he felt false if he could be of some small use to someone else.

He moved out of the playroom into a corridor of patient rooms. Ahead of him, at the end of the hall, seven dwarves turned the corner in lockstep precision and marched toward Will. After a startled second, half his mind grasped that the miniature figures were children in costume, entertainment for the daily lunch show. But a good part of him no longer looked for rational explanations.

Will stepped aside to let the dwarves pass. As he walked on, a hearty voice called out to him from behind: "Hey, Mr. S. You tryin' to sneak by here without saying hello?"

Will turned to see a big, comfortable nurse scowling ominously at him from the doorway of the chart room: Cass, a mountain of maternal instinct in the colorful pajamalike scrubs that at some undefined point nurses had started to wear instead of the starched white uniforms Will remembered from his own childhood. The big nurse's broad, dark face was Southern sunshine, sweet tea, and rollicking gospel. Will felt the knot in his chest loosen. He pressed a hand to his heart.

"Cass. How could I do that? You're the love of my life."

Cass nodded once, satisfied. "Don't you forget it."

Will's smile faded as he turned the corner.

Sobbing drifted from the open door of a room ahead of him, and though the sound was hardly unfamiliar, more the norm than the exception, he slowed as he passed, glancing in. He was startled to see the elegant man from the chapel standing at the bed of a little boy shrouded in a plastic oxygen tent. The boy's mother stood beside him, her face ravaged. She was stylishly dressed even for this notoriously expensive ward, and Will recognized her—she was a celebrity in the hospital community just as he was, though rather more rarefied: Sarah Tennyson, the lyric soprano. Will knew that she had been living at Briarwood just as he and Joanna were; that her young son had some kind of aggressive leukemia. She sobbed into her hands as she spoke with the man from the chapel, who stood quietly by her side, his back toward the door.

So she was the one he was here for. Not a husband, or Will would surely have seen him before. A friend? Relative? Another musician, or conductor, even—which would explain the European polish?

Will moved closer, drawn by the man's steady, comforting presence, and caught snatches of the singer's words, choked out between spasms of tears. Even through sobs, the voice was exquisite: "My fault . . . always my career, my tours . . . if only I hadn't . . . I never should have left him. . . ."

The litany of guilt. Will knew it too well. *If I'd been a better parent—if I'd been there, if I hadn't always been thinking of myself . . . if, if, if—*

The tall man looked toward the door, straight at Will—almost as if he had spoken aloud. Will realized to his chagrin that he had halted beside the door and was now blatantly eavesdropping.

He stepped back from the doorway and moved quickly on, walking by an Easter mural of bunnies and colored eggs painted on the yellow wall, to stop in the doorway of a twelve-by-twelve room.

The standard hospital cubicle was an explosion of color,

decorated as lavishly as a child's playroom. Joanna had managed to cram half of Sydney's bedroom into the tiny space: her favorite comforters with their oversize flowers on the bed, Maurice Sendak prints on the wall. Lacquered shelves overflowed with books, games, and a mind-boggling collection of toy rabbits, from Bugs Bunny to Tenniel.

On the high hospital bed, Joanna cuddled with Sydney, holding open a book, reading. Such a familiar scene, but twisted, out of joint, for their daughter was a wraith, savaged by her cancer and the treatment, almost unrecognizable as the sunny, confident presence she had been.

Joanna too was a stranger now, locked behind glass, remote in her grief, the rest of the world shut out as she concentrated all her being on willing Sydney well.

She turned a page of the worn book with its extravagant color plates—one of the old classic fairy tales that Sydney loved for their lush romanticism and sheer bloodiness, even at five disdaining anything simple or modern.

"The Queen went out into the forest . . . following the moon path . . ." Joanna's voice was low, spellbinding. Sydney listened, rapt.

Will stood against the doorway, watching the two of them, his heart aching. As she read, Joanna stroked Sydney's cheek, the wispy remains of her hair—a touch like sunlight, like spring rain . . . such love . . .

" 'You cannot find him,' the old woman said. *'For he lives in a castle which lies East of the Sun and West of the Moon . . .' "*

Sydney looked up from the book, straight at Will—that uncanny awareness of hers. "Daddy."

For a moment, Will felt like an intruder. He moved to the bed to kiss Sydney, hold her impossibly light body.

When he straightened, the small white rabbit was on her chest, snuffling around. Sydney squealed in delight. "Daddy!"

Will shushed her conspiratorially, glancing toward the door. Animals were, of course, strictly forbidden in the hospital, but Will no longer cared about rules or consequences. Sydney hushed immediately, stage-whispered:

"What's his name?"

Will put his hand over Sydney's, petting the bunny with her. "You tell me, Princess."

Sydney answered immediately, definitively. "White Rabbit."

Will almost smiled, remembering late nights with Jefferson Airplane on the stereo, Joanna's favorite song. He looked to her to share the joke, and her eyes touched his with their shock of dark blue. But her expression was unreadable. She leaned forward to stroke the rabbit's ears, and Will caught a whiff of her perfume, jasmine and hyacinth, and Joanna's own indefinable, intoxicating scent underneath. For a moment, he was dizzy with wanting her.

In the months since the diagnosis, Joanna had not left Sydney's room for anything more than a doctor's consult. Will had been able to take a suite for them in the tower of doctors' residences adjoining the hospital, but Joanna had never even seen it; she slept on a cot in Sydney's room, and Will could not persuade her otherwise. He suspected—no, he was certain—that Joanna believed she could keep Sydney alive by sheer vigilance and that leaving her alone for more than a few seconds, any small adjustment in routine, any forgotten ritual, a slip of her undivided attention, would result in their daughter's death.

Will was helpless in the face of her devotion. It would have been monstrous to fault her for it. But under the other constant pain and disorientation, missing her was slow torture.

Without warning, a white-coated young resident walked into the room, clipboard in hand, edgy, hollow-eyed, as scruffily brooding as Kurt Cobain under his three-day stubble: Dr. Connor (part of Sydney's "oncological team," as the hospital called it).

Connor looked straight at the bed, lasering in on the rabbit on

Sydney's chest. The kid looked half Will's age, but Will felt as busted as a high schooler caught smoking in the john.

Connor's expression never changed as he spoke sardonically. "Really not getting enough sleep. I'm hallucinating rodents in the patients' rooms. Maybe if I lie down for a while, they'll go away."

He met Will's eyes—a silent summons—and backed out of the room.

Will's stomach lurched at this new unknown. He leaned against the bed, stroked the rabbit with a finger. "We'll keep him at home for you, honey."

Sydney sighed, an infinitely old sound. "I'll never see him, then." She petted the rabbit.

The words knifed Will in the heart. He looked toward his wife. Joanna stood, looking down on their daughter. Her face was a mask . . . lovely . . . barely human.

And Will's heart broke again—with grief and longing for his beloved, his soul, who had disappeared from him that day and had never returned.

CHAPTER FOUR

The tumor is not responding to the chemotherapy."

Will and Joanna sat frozen on a deep couch in a black-leather-and-chrome office. Several doctors were seated about the room; Connor stood against the wall.

On the light board was a single X-ray film. No medical training was needed to recognize the tumor: grapefruit-size, malignant, mocking, tentacles reaching hungrily through organs it dwarfed.

It was the Enemy—pure evil, an alien life force eating their daughter's life away. Wrapped around a vein, squeezing out bladder and kidney, completely inoperable. Chemotherapy—the blasts of toxic chemicals—and radiation had been their only hope.

Dr. Mankau, a chocolate-eyed Indian pediatric oncologist whose age Will had never been able to determine, sat behind the wide desk, doing all the talking in his lilting English. Behind him was a lush and mythic tapestry of gods and animal beings in saffron and emerald and indigo, the stark room's only ornament.

"The cancer has metastasized to her adrenals. It's shown up in the marrow again."

The silence was so dense, Will could hear the digital clock

clicking softly on the wall. He spoke through a dry mouth. "I'll do another marrow transplant. Take more."

Mankau said kindly, "The last one was . . . unsuccessful. Rather than put Sydney through that kind of pain again—"

Will heard himself saying from a great distance, "No, then. No."

Through the pounding of panic in his head, he reached for Joanna's hand. She sat as if carved in ice, not responding to his touch.

Mankau cleared his throat. "With your permission, there is an experimental treatment—a combination of endostatin and angiostatin—one of the so-called smart bombs. It works by turning off the blood-supply systems to targeted tumors. It has had some success in breast cancer. It is totally untried for Sydney's type of tumor, but at this point . . ."

Will nodded, fighting for control. His voice was hoarse. "Yes. Anything . . ." He looked toward Joanna. She didn't move.

Mankau glanced obliquely toward her. "I would also like to schedule a meeting for you both at our counseling center."

Will jolted as Joanna stood abruptly, staring at Mankau as if they were alone in the room. "Why?"

Her face was glass, but what Will saw there filled him with more dread than anything that had gone before.

Mankau didn't blink. "We have many fine counselors on staff. Your daughter is not the only one who needs care right now—"

"And what do they have for us? Talk? Books? *God?*" Joanna's voice rose in raw, wrenching fury as she spun in the room, looking around at the men, her eyes wild, unfocused. "What I need? Don't tell me what *I* need. *What can you do for my daughter?*"

The doctors sat, speechless.

Joanna turned and hurled herself against the door like a trapped animal. She found the knob and twisted it, bolted from the office. The doctors looked away from Will in silent, terrible sympathy.

Will stood. "I'm sorry." He fumbled blindly for the door.

In the reception area, he pushed through the glass doors of the doctors' wing and strode after Joanna, already far down the hall in the geometric corridor, blinding white with steel track lighting. "Joanna."

Ahead of him, she broke into a run. Will sprinted after her and caught her, bringing her around to face him. She turned her head, twisting in his hands, avoiding his eyes. He tightened his grip desperately. "Don't do this. Don't shut me out—" She was fighting him now, wildly, silently struggling. He could feel her heart and his, both pounding with terror.

"No. No. *No.*" She pushed him away with all her strength, eyes blazing.

Will faltered at the intensity of her fury—and in that moment was startled to see a tall, dark silhouette at the end of the corridor, watching them. A man, quite still—the elegant man from the chapel.

Joanna broke away from Will and ran down the corridor past the dark man as if she didn't see him.

The man did no more than glance at Joanna as she passed. His eyes met Will's in sympathy and regret. Then he turned and disappeared around a corner.

Will wasn't sure how long he stood in the hall. Time seemed to have stopped entirely, and maybe it had.

He had known—*known*—since the moment of diagnosis that this would be the outcome. And still insidious hope had tricked him, lulled him. Now there could be no more pretending.

He felt a sensation of falling from a great height and knew if he gave in to it, he would never land. He pulled himself up and said aloud, "No." He willed himself to breathe, to think.

He had uncharacteristically placed himself and his family in the doctors' hands, surrendered to their expertise. It was time for him to take charge again, to stand and take care of his family, to find the love that would somehow bear them through. . . .

He felt a sudden coolness in the air—the hall seemed subtly to brighten. He turned to see a pair of saffron-robed monks, with caramel skin and black hair ascetically shaved, walking almost noiselessly in the hall toward him. Their slight bow as they passed was formal, a simple courtesy, but Will felt a wave of calm.

What had the man in the chapel said? *There is a way. . . .*

He turned to watch the sandaled monks walk unhurriedly down the hall and found himself on the verge of tears.

Yes, he thought. *Maybe . . .*

And for the first time in the long darkness, he felt a glimmer—not redemption, but perhaps an inkling of peace.

But there was one part of his old life he had to finish.

The monks had disappeared around a corner. Will took a shaky breath and forced himself to move down the hall, toward the bank of elevators at the end. He hit the call button and waited. The fluorescent lights hummed above him; his head once again buzzed with deafening, incomprehensible thoughts. Through the din, steps moved behind him, and a male voice spoke.

"Mr. District Attorney." Will turned, expecting—no, hoping—to see the man from the chapel. Instead it was a clean-cut blond man in his thirties hovering behind him. Will tensed, knowing what was coming.

"Ted Dawes, *Boston Sentinel*. How is Sydney?"

Will suppressed a wave of rage, spoke shortly. "I'm sure you have that information." He pressed the elevator button again. The reporter drew closer, insistent.

"The people of Massachusetts are beginning to wonder how Sydney's illness will affect your run for the governor's office."

The rage rose, blind and killing this time. Will turned on the reporter—

Then Cass materialized from a doorway, bearing grimly down on the reporter like an avenging angel.

"You—how did you get up here? You get out before I call security."

The reporter took a startled step back. The elevator doors opened, and Will bolted inside to the sound of the big nurse's tirade. "Who raised you? What's wrong with you—bothering good people like that?"

The doors whispered shut. Will pressed his back against the rail of the elevator and closed his eyes, letting his heart slow from the adrenaline rush of anger. He'd been a second away from attacking the reporter.

He was suddenly aware of another presence in the elevator. Beside him, at waist level, came the sound of labored breathing. Will's eyes flew open. He looked down.

A man sat in a motorized wheelchair on the other side of the elevator, with the crew cut and barrel chest of a soldier. It took a startled second for Will to fully register that the man had no arms and no legs—just a torso with stumps. The limbless man's breath rasped around the mouthpiece that controlled the wheelchair. As Will stood, frozen in pity and revulsion, the man glared up . . . his pale eyes impotent with fury.

The elevator doors opened again, and Will stepped quickly through, fleeing.

In the hall outside he leaned against the wall, shaking.

Endless. I'm in hell. Disjointed, irrational thoughts.

Or maybe I'm losing my mind.

CHAPTER FIVE

He was walking again, the path he traveled many times a day, through swinging doors, across a glass bridge connecting two hospital buildings over a narrow access street. Snow beat against the glass tube around him, a wild assault of white.

Memories crowded in—maddening, taunting. A brilliantly sunny day: the Tudor house in the woods, the garden in full bloom . . . Will pushing Sydney on a tree swing, high into the blue of the sky; Joanna weeding in the garden, breathtakingly beautiful in a short, faded dress and straw hat . . .

Then night, and he and Joanna making love on the rug by the blazing fire, their bodies sweating, shivering . . . Joanna gasping in his ear, her nails on his skin . . .

Joanna . . .

Guilt shuddered through him—that through all the grief and rage at his daughter's illness, he suffered most for the loss of his wife.

He pushed away the traitorous thoughts, pushed them down, and moved through the double doors of the glass bridge onto a mezzanine with a long escalator. He stepped onto the moving stairs and experienced that sinking, down-the-rabbit-hole feeling of being in another dimension, as the escalator descended and around him the hospital morphed into a mall.

The first time he'd made this descent, he'd thought he must have blacked out and come to in a different place.

Below him, a skylit food court was ringed by bright and busy mini fast-food counters: McDonald's, Au Bon Pain, Round Table Pizza. Doctors, nurses, and relatives sat at scattered tables separated by lush planters. Beyond the food court, the galleria continued—an alley of boutiques and shops connecting the Children's Hospital lobby with a tower of doctors' condos and an on-site hotel, where patients' families could take up residence and literally never have to step outside the hospital complex; it was as well supplied as a small city.

No need to leave—no exit or escape, either.

In truth, Will couldn't remember the last time he'd actually been home. The house in the woods was closed down tight, the cat with a neighbor, while campaign volunteers made trips back and forth over the bridge for whatever he or Joanna needed.

Will suddenly wondered with unnerving detachment if they would ever see the house again.

He tightened his grip on the escalator handrail to keep from swaying on the step and stared straight ahead of him, his eyes aching.

Suddenly, a disembodied voice whispered in his ear: *"How? . . . How?"*

Will turned on the escalator, startled.

There was no one behind him.

He blinked; his pulse jolted. He turned again, scanning the food court below.

His eyes found a blue-collar husband and wife sitting huddled over a table beside a planter. The wife whispered; the acoustics of the mall made her stricken voice as clear as if she were speaking in Will's ear:

"How are we going to pay?"

After the initial relief that he was not hearing voices, Will's gut contracted in painful empathy. Even with the best insurance available for purchase, Sydney's treatment, the parade of specialists, had decimated Will's inheritance and most of their investments. He had no idea how the average family, much less the poor, were able to cope. Despite the recent passage of Massachusetts's landmark universal-health-care bill, coverage was by no means a certainty—

You can stop thinking like a politician now, he reminded himself thinly.

He stepped off the escalator and walked on the parquet tile of the galleria, past a pharmacy, clothing stores, a Citibank, a small gym franchise, and a cocktail lounge connected to the hotel lobby. He punched the button of the elevator across from the bar. Around him, mannequins watched with frozen stares from behind glass.

Will stepped out of the elevator on the seventh floor of the doctors' residence tower, into a corridor wallpapered in subtle gold patterns. A visiting German surgeon was locking the door of his apartment across the hall. He nodded to Will as Will unlocked another apartment door—and looked through the doorway into his former life.

The suite's living room was a frenzy of motion, a babble of overlapping voices and tense energy. Two televisions were tuned to different local news stations; the walls were papered with district maps of Massachusetts bristling with color-coded pushpins. Several men and women, Will's core campaign staff, paced around a conference table in the small living room: campaign manager Jerry, chief of staff Ellen, strategist Pete, their team of young, idealistic volunteers. With fierce optimism, they'd continued the campaign, stealing time from Will where they could, even bringing potential contributors to meet him in the hospital suite. Joanna had been adamant that he continue; she insisted on as much normalcy as they could muster.

But the campaign was holding on by a thread—and that only because it was not Will's first time out. Even if he hadn't already made successful runs for the city council and for district attorney, he had seen his father in action for so long, he could campaign in his sleep, which for the last month he virtually had been. And there were powerful forces that wanted him to be elected, who kept the machine running. After twelve years of Republican governors, the party was looking to break the lock.

And yet he was losing. Will looked to one of the TVs, at the face of his opponent.

The candidate smirked out of the screen, a huge American flag behind him: the smug, jowled face of Arthur Benullo, former mayor of Boston, incumbent governor of Massachusetts.

"The values that made this country great will serve us now in the battle against our enemies . . ."

Jerry glowered at the TV. Though Will's campaign manager was past forty, professionally polished, a veteran of three successful national Senate campaigns, Will looked at him and still saw his freshman roommate: Ivy League smooth, a laid-back grin disguising the instincts of a barracuda.

"Old blowhard thinks he's got it in the bag, and he's right. We've dropped ten points in the polls since January. We should not be behind Auerbach in the goddamn primary. . . ."

Ellen said softly, "But, Jer, he's not out there. He hasn't made an appearance for a month."

Pete pushed back from the table. "We're damned if we do, damned if we don't. Voters don't like it if the man is out campaigning while his little girl is sick . . . but if he doesn't, they assume he's not up to the job."

Will stood quietly in the doorway, watching them talk about his fate—feeling so disconnected, he wondered for a moment if he was really there at all. He was tempted to step back and close the door.

But old responsibilities stirred, and he cleared his throat to announce his presence.

All activity stopped as he closed the door behind him and moved into the room.

Jerry stood, muting the TVs. The sudden quiet was ominous. "Will."

Will looked around at his staff. "I need you to draft an announcement. I'm dropping out of the race."

Stunned silence from the room.

Jerry looked around at the others, nodded tersely toward the door. The staff instantly, wordlessly, gathered papers and laptops and hurried out, a few with quick, veiled glances at Will. Ellen touched his arm in silent support as she passed. No one else could meet his eyes.

Will turned and moved past whiteboards hung on the wall, with monthly and weekly calendars marking deadlines and appointments. In the kitchenette, the little white rabbit snuffled inside a cage on the sink counter. Will opened the refrigerator and took some lettuce from a shelf, turned to the cage, and dropped leaves through the mesh.

Jerry shut the door behind the last of the staff and approached Will cautiously.

"So things . . . aren't good."

Will laughed shortly, watched the rabbit eat. "No. Not good."

In the silence, the snow swirled outside the windows. Jerry seemed to struggle with himself, then the words came in a burst.

"Will, you can't drop out. It's your year. If you leave the party with Auerbach to face off with Benullo, Benullo is back in by a landslide. He'll consider it a mandate. You know what four more years of Il Duce is going to look like."

Jerry was overstating the case for effect. Another Benullo term wouldn't be the end of the world as they knew it. Will had clashed with him for four years on the city council, when then-mayor Benullo was running Boston as his personal fiefdom. Will had been the only councilman with the political cachet to stand up to him. And oh, how the mayor had hated Will, his perfect pedigree and his reforming idealism.

Will had known men like Benullo all his life, knew him for a bully and narcissist. It had not always been the case. Benullo had been a competent, if unimaginative, councilman himself, but his personality had changed radically the year his wife had died of a degenerative muscle disease and Benullo had won the mayoral election. Now the man's oily proletarian charm seemed nothing but a thin veneer disguising misogyny, elitism, and a virulent racism.

In its bones, the Commonwealth of Massachusetts was too progressive a state for a leader like Benullo to do irreparable damage, but there would be a subtle erosion of social programs, of public education, a continuing flight of the best and brightest out of the state—and certainly no forward-thinking planning, when in this state of all states, the "thinking center of the continent," the Athens of America, the home of American giants, there was a chance, a responsibility, to do so much more. Not just for Massachusetts, but for the country . . .

Jerry mistook Will's silence for agreement and leaned forward urgently. "It's *now,* Will," he said, either consciously or unconsciously paraphrasing one of Will's own campaign slogans. "You know it's now. You can take this if you just hang in there until—"

Will swept his arm across the long table, sending files, pens, and newspapers crashing to the floor, then upended the table, flinging it away from him.

Jerry backed away, stunned. Will suddenly saw himself from out of his own body, standing above the overturned table and scattered files on the floor. He had a vague sensation that he'd caused the disarray, but he had no actual memory of doing it. He had no idea how to get back into his body.

Jerry was speaking—no, shouting, "Will . . . *Will!*"

Will snapped back to reality, felt himself inside himself again. Jerry hovered, watching him warily. He looked at Jerry helplessly, groped for words. "I . . . I'm sorry." He reached for the table to put it back upright.

"Will." The force in Jerry's voice made Will stop and look at him. "You've been living here for weeks. You can't make this decision in the state you're in. You need to get out, you need sleep, you need—"

Will slammed the table back to the floor. Jerry fell silent at the look in his eyes. Will pointed a shaky finger at him.

"I watched my father trade his family for politics. I watched it kill my mother." He took a deep, rasping breath and said aloud the thing that he had not even let himself think before just that second.

"We're going to lose Sydney. I won't lose Joanna, too."

He saw Jerry flinch back as if struck. There was a long, terrible pause as the finality of the words sank in.

Jerry swallowed. "Will, I . . ."

Will watched him falter, the words dying. What could there possibly be to say?

Jerry stepped forward and squeezed Will's shoulder. "Please. Get some rest." He pulled a coat from a chair. Will barely noticed him cross to the door and out.

Will sat heavily on the couch and put his head in his hands. He remained motionless as the evening darkened to blue outside the windows.

CHAPTER SIX

*I*n *the dark hospital room, the cardiac monitor beeps steadily, flashing pulses of red light. Sydney breathes raggedly in her sleep.*

Joanna's cot is made up in the corner of the room, unslept in.

She sits in the dark beside Sydney's bed, pale as porcelain, dark hair falling loose around her face; perfectly still, unnerving in the intensity of her concentration on her daughter.

For Joanna, night is the time of demons—terrifying shadows from her childhood, night sweats and panics.

Dark thoughts . . . dark. The dark always finds her.

She pushes all this into a corner of her mind, leans closer to where Sydney lies in the high bed. Joanna takes her daughter's hand and breathes slowly in . . . and out . . . Sydney's breath adjusts to her mother's rhythm.

A shadow falls across the child's face as someone steps into the light of the doorway.

Joanna looks up, straight at the watcher—

Will jerked awake, his eyes wide in the darkness. *"Joanna—"*

He was on the couch in the suite. It was pitch black outside the windows, snow flying again. His heart was still beating crazily, a remnant of fear from his dream.

Something coming for her.

Not Sydney.

Joanna.

He sat up, groped for the light.

CHAPTER SEVEN

The white geometric hall was a shortcut, such as it was, that Will had found one night while wandering lost through one of the other hospitals. He strode in the corridor, unable to shake the feeling of danger from his dream.

The hospital at night was its own twilight world. All lights dimmed; nurses walking soundlessly on crepe soles; sleepless patients prowling the halls in their robes. Eerie night sounds floated from the open doors of dark rooms: the electronic blips of monitors . . . labored breathing . . . a soft cry. Each doorway was a window onto someone's private hell.

A nurse exited a set of double doors, and Will got a glimpse of a long ward where burn patients lay in rows of beds, Epicel skin grafts covered by gauze, some completely wrapped, like living mummies; others with melted half-faces exposed, lidless eyes bleary with pain.

Will shivered and hurried on.

In the daylight world he could pretend to work, pretend to have a life, to have hope. But at night, the full despair of his situation surrounded him. After the dissociative episode in his suite, he felt even more that he was loose from his moorings; in withdrawing from the race, he'd cut his last tie to his normal life, to reality, and surrendered to the hospital's alternate dimension.

Underneath the sound of his own hollow footsteps, he became aware of a strange metallic clumping. He glanced down a corridor to the left and was startled to see the limbless man from the elevator. The man now teetered on skeletal metal legs, walking with a jerky, robotic gait, holding himself up on metal crutches attached to his stumps of shoulders.

Will watched the insectoid walk with horrified fascination.

The limbless man swayed and fell hard to the polished floor, crutches and prosthetics tangled in a heap.

Will jumped to help him, but to his shock, the man flinched away and flailed out at Will with his crutches.

"Get away! I won't do it! Get away!" The limbless man banged his metal crutches on the floor with an inhuman howl.

Will backed up and fled down the corridor, the man's bellows echoing behind him.

He was still off balance as he crossed one of the myriad glass bridges connecting the hospitals above the icy streets. Outside, the blizzard raged, darkening the sky beyond black, burying cars parked in the streets below.

A siren screamed, piercing the muffled silence. What was left of the DA in Will paused to look down on the snow-covered street.

Below him, an ambulance rushed up the access road to the emergency entrance of the Gothic brick structure across the narrow back street. Two patrol cars followed, cherry lights swirling, their sirens layering on top of the shriek of the ambulance.

The lights of the emergency entrance illuminated the scene with the surreal clarity of a stage set. Will could see perfectly as paramedics pulled a stretcher out of the back of the ambulance.

The stretcher bore a uniformed cop with fiery red hair. The

cop's lower abdomen was obliterated, the uniform shredded by bullets, his torso a crimson lake of blood.

A shudder ran through Will as he stared down. Wind gusted through the street, swirling icy flakes.

A very pregnant young woman with red-gold hair fumbled out of the passenger side of one of the patrol cars. She stood frozen, looking after the stretcher, her face a rictus of terror. The cop's wife, Will was sure. She swayed, her knees buckling.

Will put a hand to the glass in front of him instinctively, afraid she would fall—

And suddenly a hand caught the woman's arm, bracing her up.

Will blinked as he recognized the dark man from the chapel again, coatless in the snow, tall and elegant in his dark suit.

The cop's wife thrashed in his grasp for a moment, blind panic, but the tall man held her firmly. Will stared down, his hands pressed against the glass, as the man spoke to her, his gentleness apparent even from the distance. The woman went oddly still as he spoke to her; her eyes gradually lost the unfocused, glazed look and became intent.

She nodded once and let the man lead her into the hospital after her dying husband, leaving the snowy stage empty again.

Will pushed out of the swinging doors from the glass bridge and stood for a moment, collecting himself.

Everywhere, he thought. Four times today he'd seen the man.

Surely the tall man was a counselor, then; that's what he'd been doing with Sarah Tennyson in her son's room.

Will's prosecutor instinct overruled the thought immediately. *Too well dressed for a counselor. Too polished.* Even though the dark man had no discernible accent, Will had had more the impression of a foreign businessman.

And why had he been in the chapel the first time Will had seen him?

Will's face flushed. Had Dr. Mankau sent him specifically to talk to Will? Maybe the counselor idea was not out of the realm of possibility. A European psychiatrist? After all, surgeons from all over the world were routinely flown in on cases here. A celebrity psychotherapist, even? The man had been in the opera singer's room—

A bell dinged softly, and the elevator doors down the hall from him slid open. Will glanced toward the sound and was startled to see Sarah Tennyson herself, backlit by fluorescents, just stepping into the elevator.

Struck by the coincidence, he called ahead, "Can you hold it, please?" He broke into a run after her, aware of thinking briefly that he could now simply ask about the man.

He reached the elevator just at the last moment and grasped the rubber edge of a door with his hand. The doors slid open again, and Will squeezed inside, breathing out, "Made it—"

He stopped dead. The elevator was empty.

Will looked around him in total disbelief. His own multiple reflection stared back at him from the mirrored walls.

The doors jerked shut. After a moment, he hit the button for the cancer ward and leaned against the side of the elevator as it jolted upward. His mind was reeling.

What the hell?

Dissociation, and now hallucinations?

He'd been thinking of the singer and then just imagined he saw her?

He glanced at himself in the mirrored wall: He looked like the walking dead.

When exactly had he slept last, a real night's sleep?

He looked at the mirror again and frowned. *Strange to have mirrors in a hospital elevator, isn't it?* He couldn't remember seeing mirrors in any of the elevators before. But maybe he hadn't noticed.

The dreamlike sense of unreality continued as the elevator let him off at the Children's cancer ward and he walked through the darkened playroom, through the towering animal sculptures, the oversize alphabet blocks, feeling sickly as if he'd shrunk to child size.

His palms were sweating when he finally reached the dim yellow corridor with the Easter mural on the wall.

He took a breath and tried to let it out slowly as he walked past the dark rooms with their always open doors. A soft, flickering light came from a doorway ahead.

Inside, a lovely Italian-looking girl of perhaps eighteen knelt by the bed of a much younger boy. Votive candles shimmered on the dresser and nightstand; the girl's face was wreathed in their glow as she prayed, a rosary moving like water through her hands.

Will felt a stab to his heart. *Oh God, if only. If only I could believe—something. Anything.*

Through the wavering candlelight, he saw his childhood in flashes: shuttled back and forth between churches—the society events of his mother's upscale Episcopalian congregation and the more primitive rituals of his father's Catholicism. Will had been too able to compare and contrast, to find the holes and hypocrisies in both.

He looked at the praying girl and suddenly saw his father on his knees before the altar, taking the communion wafer on his tongue.

And for the first time, Will knew without a doubt that despite the grandstanding politics of his father's religious practice, Hal had believed, to the core of him, believed all of it: Jesus, Mary, God, and the Devil. As purely as did this girl in front of Will, kneeling in candlelight.

Will turned away from the room, back into the dimly lit hall—and froze.

At the end of the corridor, a thin, stooped male patient in a robe hovered outside Sydney's room. The patient stole a furtive look behind him, then slipped in through the doorway.

Will shouted, *"Hey!"* and ran forward, pounding past the dark mouths of doors.

He bolted through the doorway into Sydney's room and spun wildly, staring into the dark.

Sydney slept fitfully on the high hospital bed. There was no one else in the room. The empty cot was made up beside the bed. No sign of Joanna or of the thin patient Will had seen.

Will swayed on his feet with a strong wave of disorientation.

Christ, what's happening to me?

He looked at the empty cot made up beside the bed and felt another frisson of anxiety. Joanna never left Sydney alone. Never.

He moved to the bed to look down at his daughter. Her tiny hands were clenched as she fought for breath in her sleep. Will put his hand on her head, gently stroking her brow with his fingers. She breathed more regularly under his touch, her face smoothing out.

"That's my girl," he whispered.

A rattling intake of breath came from just behind him—and a hoarse, frightening voice.

"Where? . . . Where?"

Will twisted around.

No one was there.

Then a shadow loomed up in the bathroom door. Light from the corridor fell on his face, and Will started in shock.

The patient hovering in the doorway was a walking skeleton. No more than ninety pounds on his six-foot frame, bent, horribly ravaged.

Will lurched forward, demanding, "What are you doing in here?"

The cadaverous patient shuffled toward Will with glowing, dying eyes. "Where is he?"

Will advanced on the patient, blocking him from Sydney's bed. "What are you doing?"

The patient reached and clutched at Will's sweater. "I need

him." His eyes burned with dementia, and Will suddenly understood.

He took the man by his wrists, spoke gently but clearly. "You have the wrong room. This is Briarwood Children's Medical. I think you must have taken the bridge from Carver or Mercy." He turned the man to face the bed, with Sydney's tiny, sleeping form.

"There's no one here but my daughter."

The thin man turned in a wild circle, looking at the shelves full of books and toys, at the bed, at Sydney. Then for a moment he seemed to understand. His eyes focused and cleared, and in that second Will caught a glimpse of his former looks, what must have been model-perfect features.

A voice came from the hallway behind. "Gary? . . . Gar?"

Both Will and the thin patient turned toward the door. A sensitive, fine-boned young man with dark hair and eyes stood in the doorway. He moved to the thin man with concern, and Will realized that the wizened patient was no older than this young man—thirty at most.

The patient called Gary leaned into the other man's arm, seeming shaken, confused. The darker young man stroked his head tenderly. "Got lost again?"

He turned to Will as if to explain—and stopped, his eyes widening in the dim light as he took Will in for the first time.

"Oh gosh. Gary, this is *Will Sullivan*. He's going to be governor." He colored slightly. "I'm Mark . . . this is Gary. We're voting for you."

Will stepped forward, held out a hand. Mark extended his free hand; there was a bracelet on his arm, a silver version of the AIDS ribbon. They shook, but instead of releasing Will's hand, Mark gripped it harder and said impulsively, "You *have* to win. Benullo is our worst nightmare. Well . . . you know."

Will pressed Mark's hand in his and felt an unexpected stab of guilt. He did know. With the rise of religious conservatism in the country, Benullo had become more vocal about his Catholicism and in a recent speech to Christian supporters had called homosexuality an abomination.

In withdrawing from the race, Will was abandoning these young people to a man he found morally repugnant.

Mark dropped his eyes and let go of Will's hand. He glanced at Sydney in the bed, stepped back, fretting. "I'm so sorry about this."

Will smiled reassuringly. "It's no problem at all. Good to meet you, Mark." He looked directly into the thin patient's face. "Gary . . . take care."

Gary leaned on his lover's arm, and Mark led him gently out. But as they crossed the threshold, Gary surreptitiously turned to look back at Will and held an index finger to his lips in a "Don't tell" gesture.

Will's pulse spiked.

What? What the hell?

He stood looking after them.

God, what a night.

Finally he turned, stepped beside Sydney's bed again. The IV dripped clear fluid into her spindly arm. Will held his breath as she stirred. Her eyes opened and she looked up hazily.

"Daddy?"

She sat partway up, blinking around her in sleepy confusion. "Where's the man . . . ?"

Will shushed her, stroking her face. "The man's gone, baby. I'm here."

Sydney sank back onto the pillow, eyes closing, instantly asleep again. Will stayed close, stroking her head . . . but his eyes went to Joanna's empty cot, and his face tightened. He pressed the call button.

The night nurse, pert, friendly Laurie, appeared almost instantly, alert and concerned. "What is it, Mr. Sullivan?"

Will briefly explained about finding Gary in the room, and the nurse shook her head, distressed. "Oh gosh, I'm so sorry. This hospital . . . I swear I'm lost half the time myself."

"We're perfectly all right," Will assured her. "If you could just sit with Sydney for a minute . . ."

He left Laurie by Sydney's side and moved out of Sydney's room, looked both ways down the dim hall. No sign of Joanna.

Adrenaline pumping again, he started quickly along the corridor . . .

Then he froze as his eyes fell on the black glass of the wide window overlooking the hospital's garden three stories below.

A dark, feminine figure stood still in the snow, under a frozen willow tree.

Joanna.

CHAPTER EIGHT

Will pushed out through the glass doors into the garden and stared around him in the dark. The hospital buildings surrounded the courtyard like the walls of a castle. Snow was drifted around trees, weighing down branches. Ice glittered under the light that came from windows far above; it was a terrible beauty.

Cold groped him through the mesh of his sweater—the dry chill of far below freezing, lulling and deadly because it was beyond cold, quickly beyond registering.

Will forced himself forward onto what he could find of a path, feet crunching on snow atop gravel, winding through the dark, still shapes of trees and plants. Under the distorted shadows of bare branches, bronze sculptures of animals and elves peered from the snowdrifts. Will's cheeks were frozen; the night tasted metallic in his mouth.

He followed the moon path through a line of Grecian-style statues on pedestals and was eerily reminded of a scene from Sydney's beloved fairy tales—a castle courtyard with courtiers turned to stone under a witch's curse.

As he thought it, the last statue turned to watch him pass.

Will froze on the path, staring up at the blank eyes.

Shadows moved across the statue's face in the wind, but the stone figure was still.

Madness.

Will exhaled slowly, turned from the statue, and kept moving.

He stepped out of the row of statues—and saw Joanna. Wearing only a thin dress, bare-legged, she stood alone in the snow under the willow tree, beside a marble bench. She was motionless, pale as the ice, oblivious to the snow. Across the path from her, an angel statue stood on a pedestal—larger than she was, androgynous, implacable.

Will spoke through lips numb with cold. "Joanna." His voice sounded thick, muffled in the dark.

She didn't move.

"Joanna." Will strode forward. She made no acknowledgment of his approach. He reached her and took her shoulders. She stared ahead of her, unseeing.

Now more frightened than he could bear to acknowledge, he pulled her to him, held her. She was cold and stiff in his arms, completely unresponsive, her eyes staring into the shadows.

He spoke through chattering teeth. "Come inside." He thought for a moment he would have to carry her, but when he drew her onto the path, she moved with him, docile and unresisting. He led her on the path toward the glass doors of the hospital.

Then suddenly she tensed and looked back over her shoulder.

Will looked back behind them.

The angel statue stood on its pedestal, looking at them with cold marble eyes. Will thought for a crazed second that its position had changed.

He forced himself to turn away, moved Joanna gently toward the doors. She went passively, as if sleepwalking. Above them, stars pierced the ebony dome, needles of frozen light.

Back in Sydney's room, Will thanked Laurie (whose eyes flicked briefly over Joanna before she exited) and settled Joanna on the cot beside Sydney's bed. She allowed him to lay her down, light as a feather in his arms, as obedient and pliant as a child.

He stroked her black hair back from her face.

"Joanna?" he whispered.

But she was asleep instantly, if she had even been awake before.

He pulled a chair beside her and held her hand in both of his, watching her breathe, his stomach churning with dread.

A whole new nightmare had opened up. What had possessed her to go outside? A reflex to find her own garden, which had always been her haven, her refuge?

Her nonresponse was terrifying. Sleepwalking? Or some more terrible breakdown of sanity?

He had dreamed of darkness coming for her.

He sat without moving, watching both of them.

Sydney's rattling breath had slowed to match Joanna's; their chests rose in perfect tandem.

And in Joanna's sleep, a tear slid from her closed eye, glistening on her cheek.

CHAPTER NINE

The first time he saw her, he was lost.

He had been thirty-two, in his fifth year as a prosecutor, in the middle of trying a particularly vile spousal abuse case: the wife a virtual prisoner in her home, brainwashed and battered by a sociopath of a husband. The night she finally got up the courage to escape, the husband had caught her in the parking lot of the bus station and beaten her into a coma with a baseball bat. She'd lived—whatever life she'd be able to have with irreversible brain damage.

It was the kind of trial that made Will's blood boil, a clear case of good versus evil, with no gray areas, except that spousal abuse was always a gray area in a court of law, and in his relatively short career, he'd seen the most vicious monsters set free, protected by some juror's arcane and primitive idea of marriage.

On the twentieth day of the trial, the jury went into deliberations, which was always a suspension of ordinary reality for Will. It could be an hour or it could be three weeks, but in the time the jury was out, the clock stopped, life stopped, and no real-life work or thinking could be done.

It was Will's habit to escape the courthouse and the pressure cooker of his own head by walking over to Boston Common—fifty acres of park, an island of pure history in the middle of downtown. Former cow pasture, first public park in America, site of sermons and witch hangings, encampment for British troops before the battles of Lexington and Concord—and still a central congregating point for the city and its visitors. There Will could wander for hours on the paths, losing himself in the parade of life going by, watching the people and their random connections and conversations, letting himself sink under the always enthralling spell of Boston's heritage.

It had just finished raining, and the chill had cleared the twisted downtown streets of pedestrians and late lunchers. Will turned up the collar of his coat and walked on the slick sidewalks, the light, cold mist welcome after the claustrophobic heat of the courtroom.

As always, he turned off Somerset onto Beacon Street to pass by the State House. The building had been his second home for eight years during the two terms of his father's governorship, and the sight of it never failed to elicit a shiver of something like destiny. He had spent most of his childhood under the gleaming golden dome, the exact center of Boston, the virtual compass, marked in degrees, from which point all directions and distances in the city were measured. Will had almost daily run the four blocks from their house in Louisburg Square and waved his way past the State House guards; had slid down the black lacework iron railings of the grand staircase; had run in the upper halls past the portraits of all of Massachusetts's governors, skidding to a halt in the executive offices in front of his own father's painted image. And it had always seemed to his eight-, nine-, ten-year-old mind that the blank wall beside his father's portrait was just waiting for a gilt-framed portrait of him.

He wrested himself from the memory and turned away from

the gates of the State House, crossed Beacon, and entered the Common. Fog rolled along the paths; the plaza and benches were absent the usual vendors of hot dogs and patriotic souvenirs, the homeless and rappers and break-dancers hustling for money, the graceful Asian women practicing tai chi.

Will walked on deeper into the park, past the Central Burial Ground, with its black-spiked fence and tall, flat headstones, tipped with the weight of centuries. He had no destination in mind as he crossed the street into the Public Garden, as always marveling at the instant transition from the masculine order of the Common, straight lines bisecting the close-cropped lawns, to the feminine seclusion of the Public Garden and its curving paths, classic statues, luxuriant willows overhanging the lagoon with its famous swan boats and arching bridge.

And that was where he saw her.

She stood at the edge of the lagoon, tossing corn cakes to the swans gliding on the water, a dark silk coat wrapped around her but otherwise seemingly untroubled by the cold.

How is it that we know that someone is the *one*? Is it as in the Platonic myth, that the original beings were split in two by the gods and we spend our lives in longing, alone and incomplete until we meet our true other half?

However it happens, Will knew.

At his first look at her, his heart seemed to stop. She was dark— so dark—a wave of long, thick hair almost black and eyes from this distance seeming blacker, in pale creamy skin: the elemental sensuality of a Pre-Raphaelite witch. The tailored coat was sculpted to a body as lithe and graceful as a dancer's. There was a timelessness about her and the setting that made Will wonder if he were asleep and dreaming.

He stood as if frozen, watching as she drifted along the bank of the lagoon, leaning out to call to the swans, which glided in and out toward her, dipping their necks into the water for the cakes she scattered. She half knelt and held her hand still, outstretched over the water, a cake resting on her palm. One white bird drifted closer . . . Will held his breath, well aware of the notorious aggression of the birds . . . then the swan craned its long neck to nibble delicately from her hand.

Will stood, utterly enchanted. His whole skin was alive with sensation—significance, impatience, desire, longing. His eyes lasered in on her left hand, resting on her thigh, and he was flooded with relief to find her ring finger bare.

He was hardly shy with women. He had his father's easy Irish charm, his mother's aristocratic elegance, and his own sheen of palpable goodness, burnished with a prosecutor's crusader glow.

And the fact was, he was a minor celebrity in intellectual Boston, the son of a wildly popular and devilishly charismatic former governor. He'd gone through adolescence under the scrutiny of society reporters; from age sixteen he'd made "Most Eligible Bachelor" lists and been treated like a local version of Prince Harry. Normally, he would have to do no more than turn a smile on a woman for her to smile back and subtly or not so subtly signal availability.

To his credit, Will was all too aware that everything in his life to that point had been handed to him: looks, brains, grace, charm, money.

Easy, easy, easy.

Now, for the first time, he found himself completely at a loss.

The swan dipped its head to the dark woman and sailed off again. She stood and turned from the water's edge. Will stepped back, behind a tree, out of sight.

She walked up from the edge of the water, up to the path. Will waited, not breathing, until she had gone a hundred yards . . . and then did something he'd never before done in his life.

He followed her.

Feeling as ashamed and excited as a stalker, he trailed her through the garden, his breath coming shorter than the casual pace would warrant. They seemed utterly alone in the park, an oddity that only added to the sense of significance Will felt.

She exited the gates of the Public Garden onto Arlington. Will stayed far enough behind that she was almost lost in the fog, but the relative quiet of the streets made it easy to tail her. At the corner of Arlington and Boylston, she surprised him by descending into the T station.

He jogged across the street after her.

The Green Line, toward Boston U, was packed with a mostly student crowd. Will stood in the aisle, camouflaged by a group of businessmen, watching her. She sat beside a window, her head, that perfect profile, turned to look out. Will's mind was buzzing . . . it seemed impossible that the whole streetcar was not as riveted on her as he was.

She got off at the Museum Street exit. Will followed her onto the street with a growing sense of significance and wonder. He knew this street like the back of his hand, and he knew where she must be going. It was inevitable.

And just as if propelled by fate, she turned in through the gates of the Gardner Museum.

The Gardner was to the Boston art world what the State House was to its politics. The building was a Venetian palace that Boston's famed and unconventional arts patroness Isabella Gardner, described as one of the "seven wonders of Boston," had shipped piece by piece from Italy to house her private art collection. Will had spent countless hours of his childhood and teen years in the Romanesque halls with their profusion of art treasures. His mother had served on the board, and the family had attended the courtyard concerts almost weekly.

The woman Will was following bought no ticket at the gate; the guard merely nodded to her as she passed. Staying well behind her, Will fished his wallet from his pants and found his membership card, then stepped past the lions at the Gardner's entrance.

To step into the Gardner was to step into another world. On the most dismal, icy winter day, the greenhouse courtyard was an oasis of spring: a riot of exotic flowers around a Roman mosaic floor, four-story-high pink marble walls and arched glass ceiling streaming soft natural light down onto the garden with its profusion of erotic statuary, whispering fountains, and sweet, pervasive perfume of orchids.

The woman walked through the dim, hushed gallery entrance into the cloister surrounding the courtyard, her whole body softening as the sensuality of the setting worked its magic. She dropped her coat from her shoulders, and Will felt another visceral rush at the sight of her violet dress—a luscious color in silk that slipped over her curves like water. She draped her coat over her bare arm and fished something from a pocket, which she slipped around her neck: a laminated badge on a cord. Will realized, electrified, that she worked in the museum. It was perfect. She belonged here. Everything about it was right.

She paused in an arch to look out on the courtyard. Will hung back in the shadows of the cloister, riveted by the soft light on her skin.

He was in the grip of a powerful resonance now—uncanny, synchronistic. He had always assumed he would meet someone, fall in love, marry. Of course, there had always been a sense of destiny about that, as with all of his life; he wasn't looking for an ordinary woman. He had always dated lovely, smart society girls, and he'd never recognized the bland, safe sameness of them until this moment. He hadn't known that he had been waiting for a magical sign.

She had moved out of the stone arch and was gone, but Will felt no urgency now. He moved through the dark, tiled passageway surrounding the courtyard, past headless statues with missing limbs but genitals intact and prominent, into the dimmed inner rooms, replete with tapestries and gilded furniture, Venetian mirrors and intricately carved church altars, and magnificent paintings—Goya, Titian, Rubens. Occasionally, he caught a glimpse of her through the tall arches overlooking the courtyard on every floor, and the distance was as seductive as the museum itself.

He wove through a maze of carved wooden screens connecting two galleries of the second floor and came face-to-face with her in the dark.

She drew back several steps, into the Long Gallery, with its stained glass and marble columns, and confronted him, her face set and tense.

"You followed me from the Public Garden."

The wariness about her spoke volumes about what her daily life was—fending off men wherever she went. Her whole body was poised to flee if he proved dangerous.

"I'm Will Sullivan—" he began quickly, to reassure her.

"I know who you are," she cut him off. Even guarded as she was, her voice was low, thrilling.

He leapt to explain. "I'm sorry. That was unforgivable. I don't usually—stalk people. I saw you in the park and I—can't explain it. I had to know you."

She was looking at him, assessing him. God, her eyes—not black after all, but the darkest of blue; only words like azure, cerulean, would do to describe them.

"It's a strange way to try to *know* someone," she said, quietly accusing.

"I feel strange," he admitted. "All this feels strange. You—being here. I know this place. It's like home. My mother would bring me here." He looked down through the arches on the courtyard, at the Roman throne that was its centerpiece. Her eyes followed his, and he was emboldened.

"The throne was my favorite. I never could understand why I couldn't sit in it. Evidently I once made quite a scene about it."

A ghost of a smile crossed her face, and his heart beat faster.

His impulse was to touch her. He did no more than think it and she pulled back, as if knowing—but their eyes were locked in the dark . . .

. . . and then they were not alone. A woman Will recognized from various charity events as the director of acquisitions clicked into the room on Chanel pumps to match her suit.

"Joanna, it's nearly three—" She stopped still, the annoyance in her face fading as she recognized Will. "Why, Mr. Sullivan. What a pleasant surprise."

Will's inbred facility with names saved him. "It's Will, please, Celia. How very nice to see you, as always." He stepped forward and took her hand, slightly more an affectionate squeeze than a handshake.

"Forgive me for hijacking Ms. . . ." His eyes grazed the dark woman's badge. "Donnelly," he finished smoothly, his tongue savoring the Irish name. Inside his head, he was chanting, *Joanna . . . Joanna,* like some juvenile lead in a musical. "It's been so long since I've had a proper tour of the collection, and she's been kind enough to answer some of my questions."

The director knew her major donors. "We're always delighted. Take as much time as you like. Joanna will show you around."

It was not a request.

"Of course," Joanna said. "Mr. Sullivan."

He caught her hand as they started up the dim marble stairs, and the touch was like an electric shock between them; he could feel her soundless gasp. "It's Will. And you don't have to. I wish you would, but only if you want."

She looked at him in the dark, and it was all he could do not to seize her there.

"Do you always get what you want?" she asked, low.

"Almost always," he said honestly. His voice was dry as dust. His hand was still closed around hers. "I try to deserve it."

She looked at him so deeply, he felt his heart stop again. Then she withdrew her hand from his, lowered her eyes, and began the tour description Will had heard so many times before. "The history of the Gardner begins in Renaissance Florence ..."

So they walked—through marble halls, past Gothic tapestries and Impressionist masterworks, while she spun the history of the Gardner. Interspersed with the tour, they spoke of themselves, subtly probing. Will was drunk on the sound of her voice, her passion for history, life, art—for Boston. He could barely keep his eyes off her; in a building full of treasures, nothing held a candle to her.

Every tour of the Gardner ended in the Gothic Room, at Sargent's extraordinary painting of Isabella herself, titled *Woman, an Enigma*. The portrayal had cemented Isabella's place as a Boston legend, the queen of the arts. Again, the significance was dizzying. Will knew he was standing in the room with destiny. He'd wanted one thing in his life: to be governor and then more than governor, more than his father. But that he had a certain patience for, because it seemed ... inevitable. Now he realized he wanted much more than he'd ever let himself admit. Now he was faced with something not at all certain—that he knew he could not live without.

He turned to Joanna, as darkly beautiful as the painting behind her was light—

And then his beeper buzzed, and real life crashed in on him.

He knew the number without checking and met Joanna's eyes. "My jury is in."

"Well ..." She turned away but made no move to go. He stepped closer to her, looking down on the alabaster of her neck.

"Joanna." He said it for the first time. She was silent, not looking at him.

"Please have dinner with me."

She tensed, and something made him add quickly, before she could refuse:

"If I win?"

She took a breath, turned back slightly, without looking at him. "What if you lose?"

He said simply, "I can't lose."

He won. Life without possibility of parole.

He waited at the restaurant he'd named, Boston's finest.

She didn't come.

He wasn't a drinker, but he nearly drank himself into a coma that night.

He found her in the museum the next morning, like the stalker he'd turned into, caught her in the Titian Room, with its lush crimson silk walls.

She tried, just for a moment, to resist him.

"You don't want this," she told him.

Futile.

He kissed her, and nothing he had ever experienced, not losing his virginity, not his first heartbreaking crush, had prepared him for the rush of sensation and emotion of that first touch—fierce protectiveness and frenzied possessiveness and terrifying dependence. He felt like an animal and a god.

From that day on, they had never been apart.

It would be months before she told him anything of her past.

And if Joanna's father had not been long dead, Will would have killed him himself.

CHAPTER TEN

It was close to dawn when Will stepped out onto the hospital balcony, breathed in the chill of the air. He was now so shaky from lack of sleep that he had to hold on to the iron railing to stay upright.

The cold was somewhat reviving, less dangerous than it had seemed in the garden. He leaned on the rail and looked out over the vast complex of hospitals—the modern steel and glass of Fordham, ancient brick Mercy, where he'd seen the dying cop admitted . . .

Had it been just that night?

Below him was the garden with the angel statue, the bench under the willow tree.

Joanna.

He'd left them both asleep and knew he could not leave them alone for long, but he needed just a moment away, to think. . . .

He shivered—not from the cold. Joanna was slipping into some dark place. He would have to get help somehow, somewhere. But he was the only person on earth she had ever confided in, the only one she would talk to, his complicated, haunted love.

Two months into their relationship, she had shown up at his doorstep with a bottle of vodka and proceeded to get more encompassingly drunk than he'd ever seen her, before or since. That night, over the course of hours that seemed straight from hell and never to end, she told him everything about her childhood, her mother's suicide, the nightly rapes by her father, her own cutting and suicide attempts, until her escape at fifteen.

He put her to bed just before dawn and stayed beside her as she slept through the day, and when she woke in the dark of the next evening, it had been as if the previous night had never happened. She had never spoken about it since and had never allowed Will to bring it up. She had constructed her own methods of coping, long ago. Her childhood was locked away behind steel doors in her mind; she refused absolutely to revisit it with any therapist.

Unlike Will, his father had known, the first time he met Joanna. Hal's reaction, in private to Will afterward, had been shockingly brutal. "Damaged goods," he'd told Will bluntly. "She'll bring you down with her."

He was not moved by Will's outraged argument that Joanna had been a child, the innocent victim of a monstrous parent. "It's a damnable shame, but it's done. There's no repairing that kind of work. You think you can handle it, but there will come a day when you realize there's no saving her."

Hal's own hypocrisy—the fact that he'd come from mud himself, that he'd compromised every shred of natural decency he had as he clawed his way up from union worker to the city council, that he'd calculatedly married far above his own station and then proceeded quietly to drive his wife to secret, progressive alcoholism with his serial infidelities—didn't bother him in the slightest.

Will had never defied his father before. The truth was, he'd never had to. They had always been perfectly aligned in what they envisioned for Will's future. Hal was canny enough to facilitate rather than dictate; in his jovial Irish way, he'd shaped Will's political career path, smoothly opening doors wherever required, suggesting club memberships and volunteer work, athletic participation and cotillion, all the social networking and character-building activities that made for a political résumé. But he'd also taken great pride in Will's independent initiatives.

Now Will displayed a ruthlessness that showed him to be more his father's son than any conscious or unconscious mimicry of Hal that had come before. He and Joanna married immediately in a civil ceremony, and Will cut his father off completely. No contact, no mercy, no hint that he would soften.

It was the most major step he'd ever taken to define himself outside of Hal's influence. In fact, the break with his father made him as a politician. He found a new confidence and purpose. He proposed and launched a landmark sexual crimes and domestic violence unit in the DA's office, staffed by a team of experts in all related fields, and channeled his fury and desire to avenge Joanna into prosecuting sex offenders and pressing for a successful repeal of the statute of limitations on prosecuting sex crimes. Within six months, he was named deputy DA.

By then, Hal had quietly conceded, though not before he tried going to Joanna with the intent of paying her off to disappear, a secret Joanna kept from Will until after Hal's death.

It was Joanna who insisted on the reconciliation, who tranquilly agreed to the lavish society wedding Hal pressed on them, in order to wine and dine the old Boston political aristocracy and "lock in Will's constituent base."

On that day, Joanna played the perfect daughter-in-law to Hal's loving father-in-law, the consummate actress throughout.

And since that beginning, she had been the ultimate political wife; her beauty and aesthetic sense contributed to Will's aura of royalty. She was at least on the surface a bride fit for a king.

Will believed beyond the slightest doubt that he was the one and only love of her life, until Sydney was born. And a more devoted mother had never existed.

But there was a darker side to that devotion. He had found Joanna in the hospital room the night Sydney was born, holding their daughter, whispering over and over beside her head, "No one will ever hurt you. No one. No one . . ."

And now the cancer, this thing eating Sydney from inside, was destroying Joanna, too. She had been poised on a razor's edge ever since Sydney's diagnosis. Last night, she might just have slipped into the abyss.

Will suddenly heard his father's voice, as clearly as if he were standing on the balcony behind him. *You can't save her.*

Every muscle in Will's body hardened, and he said aloud to his dead father, "Oh yes, I will."

He felt a cold breeze, felt Hal hovering. Gooseflesh rose on his skin. Will steeled himself, turned slowly away from the railing—

No one.

He passed a hand over his haggard face, pressed his eyes shut. Without realizing it, he made a gesture, reaching reflexively into his breast pocket for the cigarettes he had given up ten years before.

His fingers met empty cloth, and he shook his head at himself, exhaling.

The door opened behind him and someone stepped out. Will glanced back—

—to see the tall man from the chapel.

Will stiffened, quickly pulled himself together, composing his face.

The man hesitated, then nodded at Will and moved a discreet distance away to the rail. He held a steaming cup of coffee. Will's stomach contracted at the fragrance.

He found to his surprise that he did not feel any sense of

intrusion—rather, an almost overwhelming sense of relief, even expectation.

He watched the man look over the hospital complex just as Will had and breathe in the cold air. Will noted again the elegance in the cut of his slate-gray suit, the gaunt beauty of his profile, the small, expensive details of his dress: the shimmering silk of his tie, the polish of Italian shoes, the aura of masculine confidence and sexuality. Will had grown up among men of distinction: politicians, billionaires, captains of industry; Boston was still an aristocratic bastion. Yet a unique and strangely magnetic power radiated from this man—not mere money or style, but *power*.

The man reached into his breast pocket and drew out a pack of cigarettes. He tapped one out and, as an afterthought, offered the pack to Will.

Will almost chuckled in spite of himself. "I'd love to, but I quit."

The man shook his head ruefully. "I admire willpower. In other people." He lit up, exhaled, looked out over the icy garden.

"Nights are the worst around here. People start thinking that the sun won't ever come up again."

Will was startled to find his own emotion voiced so precisely. He looked at the man, spoke through a dry mouth.

"Will it?"

The man glanced at Will. Then he nodded toward the east. At that exact moment, the sun crested the horizon . . . the first silver rays of dawn slanted over the hospital buildings.

Will felt a little jolt. "Nice trick."

The man gestured with his cigarette. "Happens every day."

He looked at Will directly, now unabashedly studying him. Will shifted on his feet.

"I'm sorry. I didn't catch your name."

"It's Salk."

Will glanced at him questioningly. Salk laughed. "No. No relation to the good doctor."

He extended his hand, and they shook. Salk's grip was firm without trying to overpower but held a second longer than a casual introduction.

Will remembered suddenly his thought that the man was a counselor and realized this was not, after all, a chance encounter. He tensed, withdrew his hand.

Then he thought of Joanna, the blankness in her eyes.

If ever there was a time for a counselor . . .

Salk glanced away, then back, and spoke in a neutral tone, without urgency. "Mr. Sullivan, I've been wanting to speak to you for some time. I think you could use someone to talk to."

Will fought a wild urge to laugh. *Talk? And say what? My daughter's going to die. My wife may be losing her mind. There's no hope . . . no hope . . .*

Salk was looking at him steadily.

"It's so beyond—talk," Will managed.

Salk hesitated, then spoke gravely, courteously. "Mr. Sullivan. In my experience, it is sometimes when things seem most dire that a door opens."

Will felt the hair on the back of his neck rise with significance. He had a sudden strong feeling of import, that something huge was about to happen. He barely breathed as the other man continued.

"This is the best hospital in the country. Without question the doctors are doing all they can for your daughter's body." Salk's voice dropped slightly, and Will unconsciously craned forward to hear. "But we are more than a collection of cells. We have access to a power that transcends the physical."

Will fought a crushing sense of disappointment, his mind already rebelling against the New Age, positive thinking. Ever since the diagnosis, well-meaning friends and acquaintances who had always before seemed perfectly rational had come forward to suggest absurd things—acupuncture, Mexican clinics, crystals, energy therapy, apricot pits.

As if meditation could cure a tumor.

The man—Salk—seemed to sense his withdrawal and nodded. "You've heard it all before, haven't you. This is more useless noise to you. Talk is cheap."

He hesitated. "Will you come with me? There's something I'd like you to see."

Will glanced back at the doors with a sudden stab of worry. *Joanna.*

"I have to get back," he started.

"I'll walk with you," Salk offered. "It's on your way."

His voice was compelling without being overbearing, and after a moment Will nodded, without really understanding why.

Salk crushed his cigarette into the sand-filled dispenser. He opened the door for Will, and they walked, Salk negotiating the maze of still almost deserted corridors effortlessly. He did not speak again, but Will found the silence between them peaceful, strangely comforting.

They passed the playroom, with its hanging butterflies and tumbled blocks, empty and in shadows at this hour. Will cleared his throat.

"How long have you been with the hospital?"

"Forever." Salk half smiled, shrugged. "Where else? Boston is the medical capital of the world, and this is the hub of the Hub— the most amazing collection of specialists. History-making procedures are developed here almost every day."

A door swung open as they passed, and a nurse exited. Will got a brief glimpse of the MRI room with its space-age equipment. A technician slid a patient on the table into the coffinlike tube, and Will felt a chill; the sight was altogether too reminiscent of a modern torture chamber.

Salk had seen as well. He mused, almost to himself, "Incredible, isn't it, that two hundred years ago, the concept that simple cleanliness could prevent infection was virtually unknown. That for years medical students experimented with ether for their own pleasure, never guessing at a more revolutionary medical use, until William Morton thought to try anesthetizing patients during surgery."

Will looked at Salk, almost unconsciously registering the statements as perfect sound bites to work into a speech. But again he felt an inexorable pull—that Salk was working toward something of tremendous consequence.

Salk was watching him. "Imagine what life-changing breakthroughs are around the corner, just beyond our grasp."

They'd turned into a hallway that seemed familiar to Will, though he'd often stride confidently through a ward only to find himself hopelessly lost. Then he recognized the corridor around the corner from Sydney's room, where he'd seen Salk with Sarah Tennyson.

"Yes, Sarah Tennyson," Salk said beside him, and Will wondered if he'd spoken aloud. He remembered suddenly that Salk had caught him eavesdropping in the doorway, but Salk seemed to be unaware of his embarrassment. The counselor stopped outside a room a few steps back, but with a view inside. Will looked in through the doorway.

Inside, the singer was packing a suitcase with the help of a little boy dressed in street clothes. The bed was neatly made up, the oxygen tent gone. Sarah was radiant, almost unrecognizable as the desperate, weeping woman Will had seen before.

Will's mind tried to process what he was seeing. Beside him in the corridor, Salk spoke quietly. "They're going home. Anthony's leukemia is in remission."

Will stared at the other man, stunned. He'd seen the boy in an oxygen tent just . . . yesterday? The day before? The exact timing eluded him, but surely it was just a matter of a day or so.

He looked hungrily in at the tableau: mother and child, the singer's joy. He felt a million miles away from them.

Salk spoke low, beside him. "Mr. Sullivan, it happens. Every day."

Will found himself flooded with conflicting emotions—wariness, fascination, desperate yearning. He forced his eyes away from the scene to look at Salk, clinging to his words like a drowning man.

"But . . . how—"

"Don't." Salk's voice was sharp, startling. "Don't concern yourself with *how.*"

The hairs on Will's arms stood up. He felt caught in some kind of undertow, seductive and threatening. Salk's low voice was thrilling in his ear.

"The first step is merely to allow for the possibility. Without conditions or expectations, allow for the possibility." The dark man's eyes met Will's, and he spoke softly. "If one miracle has ever happened in the world, why not for you?"

The thought was so simple and startling that just for a moment, it seemed possible.

Then both men caught sight of a girl walking the hall. Her face was dazed and pale, and though she held herself erect, she walked not quite in a straight line. The beads of a silver rosary slipped through her fingers; her lips moved soundlessly. The Italian girl Will had seen praying in her brother's room.

Salk drew back from Will; his face turned grave, regretful.

"Teresa Marinaro. Her brother's had a bad turn." He shook his head, already moving to step after the girl.

Then the counselor hesitated, turned back to Will.

"She hasn't left you, you know. Believe that."

Will stood frozen, unable to breathe . . . yet feeling as if an anvil had been lifted from his chest.

Salk straightened and shrugged, breaking the hypnotic contact.

"I'm always here." He smiled wryly. "At least, so it seems." He held Will's eyes. "When you're ready. *You must never give up hope.*"

He nodded to Will and started after the girl.

Will had to suppress an urge to call Salk back as he watched him go.

CHAPTER ELEVEN

Will stood in the doorway of Sydney's room, watching his women sleep: Joanna on the cot, her hand resting on the side of Sydney's mattress, attuned to her every movement; Sydney turned in her slumber to her mother, fingers curled toward her.

He moved to the bed and looked down on them . . . Joanna looking for all the world like the color plate of Snow White in Sydney's fairy book: skin as white as snow, lips as red as blood, hair as black as ebony.

And just as surely under a spell from which she might never awake.

He tucked her hair back from her face, took her hand in his. He ached for her in every cell of his being.

She hasn't left you, you know.

Tears swam in his eyes, and he brushed at his face, took a shuddering breath.

Should he take her forcibly to the counseling center? Confide in Salk, demand medication, observation—even hospitalization?

For a black moment he wished that Sydney had never been born and in the same second was suffused with shame and self-loathing.

I'm the one who should die. I'm not worthy.

His battered mind turned in exhaustion to the elegant counselor.

Too elegant, he thought again.

But there was something undeniably comforting in Salk's presence. A sense of possibility. Even . . . redemption.

If one miracle has ever happened in the world . . .

Through the haze of sleeplessness, his waking sleep, Will saw Sarah Tennyson, her happiness lighting her son's room.

Emotions roiled in him: jealousy, rage, despair . . .

Possibility?

If one miracle has ever happened in the world, why not for you?

Anything wrong with Joanna would be healed instantly by Sydney's recovery. Will knew that as surely as he breathed.

As if hearing his thoughts, Sydney stirred on the bed, opened her eyes. She looked around the room, frowned toward Will.

Will held a finger to his lips, gestured toward Joanna. Sydney nodded, instantly understanding. Will moved quietly to sit beside the bed, reached through the railing to hold Sydney's hand, pretending for a moment to chew on her fingers and growling in a Cookie Monster voice:

"Mmm. Breakfast."

She giggled, but completely silently. They spoke in whispers.

"Morning, sweet pea. How's my girl?"

Sydney looked at him seriously. "They're going to bomb me today."

Will blinked. "What?"

"With the smart drugs. Will it hurt?"

Will used every ounce of strength he had left not to let his face change. *You bastards. Leave my little girl alone. Let me die instead.*

He stroked Sydney's nose with a finger. "Mommy and I will be right there with you the whole time, baby."

Sydney sighed, tried to smile. "Can I have ice cream for breakfast?"

He knew the request was a ruse—fake blackmail. She had no appetite at all anymore, not even for the things she'd most loved. But he played the game.

"As much as you want. Strawberry?"

"Uh-huh."

He kissed her forehead, closing his eyes. "I'll buy out the store."

He stood and moved for the door. A small voice came from behind.

"Daddy, am I going to die?"

Will's heart wrenched. He turned and looked at his daughter, and then suddenly Salk's voice spoke unbidden in his head.

You must never give up hope.

Will's mind protested, struggling.

But I can't lie to her.

Sydney's clear gray eyes were on his, waiting.

Before Will could answer, Joanna sat up from her cot. "No, baby. You're not going to die."

Her voice was hard, but normal. She stood from the cot and leaned over the rail of the bed to hold her daughter. Will watched her, not breathing. Though she avoided his eyes, she touched Sydney with tenderness and attention; there seemed nothing bizarre or disconnected about her behavior.

He tried a question. "I was going to the café. Can I bring you something?"

Joanna smoothed Sydney's hair. "What would you like, baby?"

"I'm having ice cream," Sydney said decisively.

Joanna's smile trembled. "Ice cream, huh?" Her eyes touched Will's briefly, and Will was vastly relieved to see lucidity there. "I guess that's what happens when I sleep late. You two get together and plot against me."

Her voice was normal, even wry. Joanna was always an extraordinary actress when it came to putting on a game face in front of Sydney, but she seemed so clear and connected this morning that Will wondered for a surreal moment if the night before had been a dream.

He felt a flicker of something almost like . . .

Never give up hope.

He tried to speak lightly. "I'll be right back, then." Joanna nodded, not quite looking at him.

Will stepped outside the room but watched from the door without them knowing for a moment. Joanna spoke softly to Sydney, and she giggled . . . nothing at all amiss.

Will felt light-headed as he started across the reception room, with its pale yellow walls and bright splashes of furniture.

A man rose from one of the couches.

Salk?

Then Will recognized Dawes, the *Boston Sentinel* reporter, striding toward him determinedly, tape recorder in hand.

"Mr. Sullivan, are you ready to confirm that you're dropping out of the governor's race?"

Will stopped and stood for a moment, a million miles away. The voice in his head spoke again, clear and insistent.

If one miracle has ever happened in the world . . .

"Mr. Sullivan?"

The reporter's query brought him back. Dawes hovered in front of him, unsure.

Will looked the reporter in the eyes. His words came out of the blue, shocking him.

"No. I'm not giving up."

For a split second, Dawes looked as startled as Will felt. Then the reporter regained his composure and pressed harder. "Does that mean Sydney is better?"

Will took a breath, said softly, "It means—I have hope."

Before he turned away, he saw the cynicism and pity in the reporter's face.

But for a moment, there was also something very like compassion.

Will walked past him, into an elevator, exhausted . . . but for a solitary moment, at peace.

CHAPTER TWELVE

The lights of the hospital galleria were too bright around him as he stood at the counter of the café, buying strawberry ice cream, croissants, and two mega-lattés. He swayed with exhaustion, trying to focus. Now that he was alone again, the momentary flash of optimism had disappeared and he felt sick with apprehension.

His daughter's small voice haunted him.

They're going to bomb me today.

He felt a sudden wild impulse to snatch Sydney up and flee this prison of a hospital, to take her and Joanna—

Where?

Home?

Home to die in some sort of relative peace?

To watch Sydney die, and Joanna along with her?

There was no "home," no "safe," anymore. Surely the best course was to stay, to attempt the experimental treatment.

Other voices buzzed in his head now. *A combination of endostatin and angiostatin. It works by turning off the blood-supply systems to targeted tumors. . . . This hospital is the best in the world.*

If there was a miracle to be had, it would be here.

And yet—the overpowering urge to flee. Why? A normal reaction to overwhelming stress?

Will forced himself to breathe, to focus, to categorize the flight impulse.

Danger.

That was what he had woken up with the night before. The feeling of danger was strong, and inexplicable. Death, despair, depression—all of those were to be expected, certainly, but danger?

He flashed on Joanna in the snow, mute and unseeing.

What had happened last night?

Had anything happened last night?

Could Joanna have had a trancelike episode in the night and be completely normal in the morning?

His mind fluttered through bizarre images: the limbless man on crutches, the skeletal patient and his lover, the dead cop and his pregnant wife, Sarah Tennyson disappearing in the elevator, the statue turning to look at him in the icy garden . . .

And his out-of-body experience in the hotel suite, Jerry backing away from him as if he'd been possessed . . .

The thought occurred again: Was he the one who was losing focus and sanity?

He forced down a wave of nausea at the thought. But he knew he was not functioning normally. He'd by now passed into a stage that had plagued him since law school, when he would stay up cramming for so long that his body rebelled and in some kind of masochistic revenge simply refused to let him sleep. Valerian, melatonin, Ambien—he'd tried them all with no success. And yes, he recognized the dissociation that came with extreme sleeplessness.

He felt another wave of disorientation. Had anything he remembered actually happened last night? Was it possible he'd slept and dreamed everything in a state of unconsciousness so profound, he could not distinguish it from waking?

He passed a hand over his face, squeezed his eyes shut, then opened them. Everything around him seemed heightened: the painfully bright lights in the galleria, the sharp, cold edge of the counter, the crackle of the bag as he took it from the cashier, the deafening clink of money changing hands. His own voice was slow to the point of madness as he croaked out his thanks.

No. This was dangerously far from normal.

He removed the lid of one of the tall paper cups, took a deep swallow of burning coffee, and another—and was momentarily revived by the pain and the ragged bitterness. He turned and began his walk through the hospital again, forcing himself to focus on the details of reality: the gift shop with its profusion of stuffed animals in the window and fragrance of flowers, the upward pull of the escalator, the blinding white of the glass bridge to Children's. He kept sipping coffee and felt the caffeine start to spike.

More voices blurred in his head—Mankau and the doctors in the office: *An experimental treatment . . . totally untried for Sydney's type of tumor, but at this point . . .*

At this point, what was there to lose?

Only everything.

But maybe . . . maybe . . .

You must never give up hope.

A siren wailed below as an ambulance raced up to the brick hospital on the street opposite. Will slowed, looking down through the glass of the bridge, remembering the surreal episode of the night before—the wounded cop, bleeding beyond hope of recovery, the pregnant wife who must surely now be a widow.

And Salk, coatless in the snow, saying something that brought the young wife out of her trance into that cold and disturbing focus Will had seen.

He stared out the glass and saw the scene again, with the photographic replay that made him a great prosecutor: the wife, on the edge of collapse; Salk holding her up, speaking low beside her;

and the sudden change in the wife, the deep stillness, the—yes, hardness, in her eyes—

Will was jolted back to the present with a sudden cold certainty. He hadn't dreamed it. It had happened.

What had the counselor said to her to bring on that change?

Will made a turn down an intersecting bridge and headed for the hospital across the way.

Mercy was a much older hospital than the others in the complex, still its original brick, almost churchlike with its arched doorways and high ceilings. Will had wandered into it by mistake a few times before and each time had experienced a queasy flashback to the Masses of his childhood.

The nurse at the scrolled mahogany reception desk recognized him; he could see it in her startled look as he leaned in to the counter.

"A police officer was shot down last night. I'd like to get in touch with his wife."

The nurse knew immediately. "Mrs. O'Keefe? Room 212. Downstairs and follow the blue line."

Will followed the nurse's instructions automatically, starting for the stairwell without asking why the young widow was still in the hospital. Now, headed downward with his own footsteps echoing around him, he wondered uneasily if she'd gone into premature labor or collapsed from shock. In either case, the last thing she'd want would be a sympathy call from the DA.

And truthfully, though he had paid more than a few such gut-wrenching calls on the grieving widows of fallen officers, he had his own agenda now. Much as he wanted to find out more about Salk, he knew it was completely inappropriate to approach the cop's wife in her bereavement.

He came out of the stairwell on a lower floor and looked

around for the blue line, though he was leaning toward abandoning his mission.

Someone shouted out a string of curses, and Will turned, startled.

In the hall ahead of him, two orderlies herded a group of a dozen Alzheimer's patients, vacant-eyed, stumbling, some muttering in dementia . . . an unnerving sight.

A wizened old man with a nest of tangled hair and spittle on his chin fixed a brimstone stare on Will and shouted, arm extended accusingly: "Behold a pale horse: and his name that sat on him was Death, and Hell followed with him!"

At that moment, an elevator opened behind Will. He ducked inside.

The elevator was empty. He hit a button at random, willing the doors closed, with an irrational fear that the Alzheimer's patients would crowd into the elevator with him. He could hear the scuffling footsteps approaching outside, the demented ranting: "Be sober! Be vigilant! Your adversary the Devil as a roaring lion walketh about, seeking whom he may devour—"

Then the doors slid shut. Will breathed out as the cab started down, and he noticed he was again in the incongruously mirrored elevator. Multiple images of himself crowded in around him.

As he stared into his own reflection, the elevator suddenly stopped—between floors.

Will pushed the button again. The doors remained closed. The elevator didn't move.

Will punched the button again, raw-nerved.

Then the wall in the back of the elevator slid open behind him.

Will felt the hair on the back of his neck rise. He turned . . . and looked through the back of the elevator into a dim hall.

He stepped cautiously out of the back of the elevator, looked around him.

The corridor was brick, windowless, with high arched ceilings—completely unfamiliar. Halls branched out in all directions with Escher-like perspective.

He turned back to the elevator just as the doors shut in his face. The elevator whirred upward, leaving him. Will searched the wall. There was no call button.

Through the strangeness, part of his mind registered with relief that he hadn't hallucinated Sarah Tennyson disappearing, that she and now he had somehow discovered access to an inner part of the hospital.

Will looked both ways, then started down the corridor to the left and turned the corner.

The hall he found himself in was sepia-toned, with high ceilings and an arched window at the end that illuminated the passage in an old, almost ethereal light.

Will slowed, looking ahead. He was face-to-face with a stone statue—an oversize angel on a pedestal, holding up a set of scales. It seemed oddly medieval for a hospital.

At the other end of the hall, incongruously, a grandfather clock ticked, pendulum swinging. An antique armchair sat beside it as if waiting for someone.

Will stood for a moment, thrown by the decor. He looked down one side of the hall to the other . . . and caught a glimpse into the patient room behind the angel with the balance.

Salk sat beside the hospital bed, holding the hand of a dreadfully thin patient.

Will felt a jolt as he recognized the counselor—and another as he realized the man in the bed was Gary, the AIDS patient he'd caught wandering in Sydney's room.

Salk spoke to Gary, intimately, reassuringly. Gary clutched his hand like a drowning man.

Will watched for a moment.

Surely he's not telling this *man that there's hope.*

He was instantly ashamed of his own cynicism. He stepped closer to the door, drawn by Salk's indistinct murmuring, once more struck by the radiant sense of comfort that surrounded the counselor.

He wondered again if his encounter with Salk on the balcony had not been an accident, if Mankau had sent the counselor to him specifically.

We have many fine counselors on staff . . .

But to what end, exactly? Did Salk know about the new treatment Sydney was starting today? Will felt a flutter of painful hope. Could Salk have some specific knowledge of the treatment's efficacy that he'd been about to impart, before Will had shut him down with his knee-jerk reaction to alternative healing?

He knew he had to get back to Sydney's room, but he found himself torn, keen to hear more of what Salk had started to say.

If one miracle has ever happened in the world . . .

Finally, not wanting to be caught eavesdropping again, he turned from Gary's room—

—and caught movement at the edge of his vision. At the end of the hall, a young man turned the corner beside the grandfather clock, walking toward Will. Will recognized Gary's sensitive, dark-haired lover. He was pale, haggard, and in a daze that Will found achingly familiar. He didn't even notice Will until Will spoke his name.

"Mark."

The young man stopped. He was so wrapped up in his own pain, it took him a moment to focus enough to place Will.

"Oh . . . oh, hello." Then his words came in a rush of unease. "Listen, I'm so sorry about last night. Gary's started to wander. He doesn't know what he's doing—"

Will held up a hand to reassure him. "Not a problem." He glanced down the hall toward Gary's room and was aware that it was the investigator in him who spoke next, casually. "A counselor's with him." He watched Mark for a response, wondered briefly what exactly he was testing.

Mark lit up slightly. "Oh, good. The center's great, isn't it? They really . . . try."

There was a tremor in his voice. Will looked at the younger man, asked gently, "How is he?"

"Not—good." Mark broke down, dissolving into racking, ugly sobs. Will reached out, put a hand on Mark's shoulder while he cried. The grandfather clock ticked behind them.

Mark swiped at his eyes with his sleeve. "It's all a nightmare. I don't know if I'm ever going to wake up. I think every day, I'm really losing my mind."

Will breathed in shallowly. "I know the feeling."

"It just—it turns you into an animal. Both of us . . ." Mark hesitated, and his voice dropped. "He hates me for not being sick."

Will had been a prosecutor too long not to know truth when he heard it, but he kept his face neutral. "I'm sure he doesn't."

Mark smiled sadly. But he squared his shoulders and brushed away the remnants of tears, game face on, before he walked briskly toward Gary's room.

Will was suddenly aware of the takeout bag he clutched in his hand, with Sydney's ice cream and Joanna's croissants. What was he doing? He should be with them.

He started walking again, past open doors.

Another flight of stairs later, he found himself walking in a white corridor with crossbeams like huge X's in the windows, without quite knowing how he'd gotten there—almost as if the sepia-toned hall had been a separate dimension or simply his imagination. Worse, he had no idea how long he'd spent lost in the hospital, and Sydney was scheduled for treatment any minute.

As if in reproach, the beeper he always carried vibrated in his pocket. And as always, Will's pulse skyrocketed, though he knew the vibration, as opposed to a beep, meant only a summons, not an emergency.

His pace quickened, but he had no idea where he was going. He turned another corner—

Three nuns in full habit were walking toward him, long black robes flowing, their faces shadowed by cowls. Will felt a flash of childhood trepidation, the uneasy mystery of the sisters. He pushed down the feeling and moved toward them.

"Excuse me, I'm looking for the Children's cancer ward?"

One of the nuns raised her head to look at him.

Will froze. He was staring into a grotesque visage like something carved out of knotted wood, with glaring coal-black eyes . . .

He jolted back in horror, screaming inside his mind, his head snapping back so hard that his teeth crunched with a sickening clack.

The sensation brought him back into focus, and he stared in confusion. The face in the cowl was only the wrinkled face of an elderly nun.

She regarded him sweetly. "You're in Mercy, dear. Children's is across the way. Follow the blue line back to the elevator."

The nuns moved on, gliding down the corridor like big black birds.

Will stood in a daze.

Hallucinations.

Again.

Sarah Tennyson disappearing in the elevator, the statue turning to look at him, now the nun.

Panic was rising, threatening to overwhelm him. Was it true, then? Was he slipping off the edge of sanity just as surely as Joanna?

Focus, he ordered himself. *Get back to the room.*

He forced himself to glance around and saw through his disorientation that the nun—

monster

—was right. He had picked up the blue line on the floor again.

He moved to follow it, through the blinding white corridor, past the X's in the windows. Ahead of him, a woman stepped out of a patient room. She was strawberry-haired, very young and very pregnant: the wife of the bullet-riddled cop.

Without thinking, Will spoke her name. "Mrs. O'Keefe?"

The young woman—startlingly young, barely in her twenties, he thought—stopped in her tracks, took a step back, looking at him warily, framed against the white cross in the window. The stark winter light rendered her two-dimensional, like a medieval Madonna.

Will stepped back himself. The words came automatically. "I'm Will Sullivan. I wanted to say how truly sorry I am for your loss."

There was a flicker in the young wife's eyes. Grief? Confusion? Recognition? (For a cop's wife would surely know the DA.)

But what Will actually saw, or thought he saw, was fear.

He continued, less sure. "I saw your husband brought in last night."

Now the look she gave Will with those pale blue eyes was unmistakable: hostility.

Will faltered, "I . . . only wanted—"

Then he stopped, staring in dazed shock—as the red-haired cop stepped out of the room behind her.

He was dressed in a hospital gown, chest swathed in bandages, but undeniably, miraculously alive—and sounding every bit the husband as he grumbled to his wife:

"And tell the nurse about the TV reception. I can't get the game."

Will stared toward the cop's wife in sheer disbelief. She gave Will a strange look . . . *yes, frightened* . . . then took her husband's arm.

"You get back into bed right now."

She steered him into the room, scolding, as Will stood in the corridor, stupefied.

CHAPTER THIRTEEN

Somehow he was back on the Children's ward, walking out of the elevator into reception, with its big colorful couches and child-size play tables and circular nurses' station that reached out like rounded arms.

Dr. Connor, the edgy young resident from Sydney's team, stood at the counter writing in a chart. He glanced at Will. "There you are. Good. We're starting at nine." He continued writing.

Will did not respond. He stood in the middle of the round blue carpet, reeling with this new fear. At what point did the line tethering one to sanity snap altogether?

If he'd seen a monstrous nun, were the live cop and his Madonna-like wife just another hallucination as well? It was almost more likely than the alternative.

In his mind, he saw the flurry of snow around the ambulance, the officer's ruined uniform, ripped to shreds, his torso a lake of blood—

That cop was dead.

Will was still clutching the bakery bag. He put it on the counter gingerly, feeling the crackle of paper under his hand. The scent of chocolate was rich and real. He raised his latte, gulped coffee, and felt more grounded. He reached into the bag, took out a croissant, and offered it to Connor.

The young doctor's face lit up like a twelve-year-old's. "Thanks. Can't remember when I ate last."

As Connor devoured the pastry, Will took another swallow of coffee, looked off into space. He spoke carefully. "There was a cop brought in last night with multiple gunshot wounds. Do you know anything about that?"

Connor frowned at him over the croissant. "Brought in *here*?"

"The hospital across the way. Mercy."

Connor's face cleared. He shrugged. "All six hospitals have separate admissions. A thousand patients come through every day. I can barely keep track of my own caseload."

The expression on Will's face evidently mirrored the turmoil inside him. Connor raised an eyebrow. "What about it?"

Will kept his voice neutral, professional—just the facts. "I've been a prosecutor for fifteen years. I've seen a hundred gunshot wounds. I would have bet money that officer was dead. But he was walking around this morning like nothing happened."

Connor gave him a tolerant look. "It *is* our job. Hospital, you know."

Miracles? Will thought, his face heating. But the kid sounded so blasé.

Will nodded. "Right. Right." A voice in his head whispered back, *And what about the nun?*

He saw again the blistered bark face, the anthracite eyes under the cowl.

Connor brushed crumbs from his hands and picked up his chart. "Cass is prepping Sydney for the first round of angiostatin. Let's do this."

The words were like ice water. Will was suffused with guilt.

What was he doing, wandering around chasing phantoms instead of being with her?

Connor touched his arm, bringing him back to the present. The young doctor was looking at him with calm support. "New day, right? You never know."

Will turned with Connor toward the room, leaving the bakery bag with Sydney's melted ice cream forgotten on the counter.

The infusions were a nightmare; they always were. Straight shots of poison into a little body that had already borne more than any child should have to endure. Worse than the punctures, the racking nausea, was the absolute stoicism with which Sydney took the treatments. Nothing but silent tears while her whole body fought the intrusion of the tubes and toxins.

At least, whatever darkness had gripped Joanna in the night, she was now as present and lucid as anyone could be. She sat in the bed with Sydney, holding her cradled on her lap, their bodies almost fused together, while she sang softly into her hair.

Will stood by, helpless in the agony of watching his child suffer. The kindness of the doctors, Connor's and Cass's jokes, their obvious love for Sydney, were in their own way unbearable.

How could he have thought there was a chance?

He was paralyzed with fear and rage, screaming inside his head . . .

Then he felt a presence, had a distinct feeling that he was not alone.

He knew before he turned to the door what he would see.

Salk stood in the doorway, framed in the light from the hall, standing still and radiant, watching with steady compassion.

His eyes met Will's.

Will almost stepped toward the door—then Sydney spoke tearfully behind him. "Daddy, my mouth is hot. . . ." Will turned from the door and went to the bed, reached for the cup of crushed ice on the bedstand. Joanna wrapped her arms around Sydney from behind, her cheek against Sydney's head, while Will fed her ice chips.

Beside the IV stand, Cass watched Will with troubled eyes. As he kissed Sydney's hand, shushing her, the nurse suddenly said in a wondering singsong, "Lord, I have never seen such a fighting child. You a wrestler, baby? One of those big ol' Mack truck sumo wrestlers on TV?"

In spite of everything, Sydney giggled weakly at the absurdity. *"Noooo."*

Will looked at Cass gratefully. Cass studied him, then she nodded once, looking reassured. She turned back to Sydney.

"You sure about that? 'Cause you're so strong, I'm thinking I got me a sumo wrestler in disguise."

When Will looked back at the door, Salk was gone.

Later, much later, the three of them slept, passed out in exhaustion: Sydney in the hospital bed, IV rehydrating her after hours of vomiting; Joanna in the cot beside her, clutching her hand; Will in an armchair beside both of them.

A tall, dark shadow moved soundlessly into the room with them, approaching the bed.

Will jerked awake—

—and saw Cass leaning over him. She whispered to him, "You go back to your room and lie down straight out. I don't need you in traction on top of everything else."

Will had to smile at her gruff humor. He rotated his neck gingerly; she was right that he was about an hour away from an orthopedic emergency.

He glanced immediately toward Joanna and Sydney, ashamed. His twinge of discomfort didn't even register on the scale.

"Go," Cass ordered. "I'll buzz you if they need you."

Will nodded, rubbing his face, but his eyes remained on Joanna as she slept. He spoke suddenly, softly.

"Cass, I don't know what to do. I'm trying to have faith. . . ."

Cass's eyes sparked in the dark. "You have plenty of faith. Don't you let anyone tell you you don't."

Will was startled into silence. The big nurse's gaze drifted to Joanna, and her eyes seemed to flicker with unease. But if she was going to say anything, she thought better of it. She turned back to Will earnestly.

"Mr. S., I don't know where your belief is, but it's all God. It's all love. All kinds of paths open up to people in fear. You stay straight and trust your heart."

Will walked in the gleaming, translucent corridors, shaky from being wrenched from the beginnings of sleep and haunted by the troubled look in Cass's eyes. Despite her clear reluctance to interfere, the worry had been stark on her face. Worry—for Joanna?

Or was he projecting that, finding a mirror of his own fears?

He rounded a corner and realized he was in the white hallway with the X-shaped crossbeams in the windows. With no conscious intent, he'd come straight back to the cop's room.

He stopped outside and looked in on the bed, where the red-haired officer slept, snoring softly.

And a voice in Will's head, the no-bullshit instinct that made him a great prosecutor, said very clearly, *This guy was dead.*

All his alarm bells were going off. The sense of not-rightness was impossible to ignore.

Will turned away from the cop's door and walked deliberately down the hall toward the chart room.

He stopped a few steps outside the closetlike room. A nurse was inside, her back to him as she sorted medications from a cart into a cabinet with cubbyholes labeled for each patient.

Will looked quickly over the room, focused on a floor-to-ceiling metal rack that held the thick patient medical charts. He scanned the alphabetical names. Halfway down the rack, he spotted the cop's name: O'Keefe.

The nurse turned, and Will stepped back from the doorway, out of sight. He peered in just far enough to see the nurse wheel the medication cart into the pharmacy closet.

As soon as she disappeared into the closet, Will took four silent strides to the rack and removed the cop's chart. He darted out the door just as the nurse stepped back into the room.

In the hall, Will looked down at the chart in his hands and felt a qualm, understanding just what an invasion it was.

Then he started grimly down the hall.

He found Dr. Connor in the dimmed staff lounge of the Children's ward, sprawled on a sofa beside a floor lamp, stubbled and bleary and swilling coffee as he went through his own patient charts. Will was reminded again of a scruffy rock star.

He put O'Keefe's chart on the table in front of the young doctor. Connor looked up at him, red-eyed.

"O'Keefe's chart," Will said brusquely. "The cop I told you about."

Connor gaped at him, aghast. "Jesus Christ. You *stole* his chart?"

Will smiled thinly. "I borrowed it."

The resident looked seriously disturbed. "Sullivan—*Mr. Sullivan*—DA or not, medical records are confidential—"

Will cut him off. "There's something wrong here."

Connor shook his head but opened the chart, glanced over it. Will watched the young doctor's face as he frowned, read aloud.

"Admitted two nights ago, multiple gunshot wounds, perforated spleen and liver. Surgeon . . ."

He registered the name, looked up. "Anton Nazaroff. Well,

that's it—your guy lucked out. They had the best transplant surgeon on the planet here that night, just finishing up on another liver. Nazaroff did a double transplant, literally rebuilt the guy from the inside out."

"There just happened to be a spleen and a liver available?" Will asked pointedly.

Connor hesitated, then lowered his voice, though they were alone in the room. "You didn't hear this from me, but cops get priority. Unwritten rule. Especially with this kind of line-of-duty injury." He looked at Will, and his eyes were surprisingly cynical. "You're a DA. Tell me your office doesn't bend over backwards—"

Will exploded in frustration, slamming his hands against the wall. "This man had a double organ transplant and he was walking around the *next day*? What do you make of that?"

Connor straightened on the the couch, watching Will warily. "I think it's stupid as hell, but it's just like a cop."

Will paced the room with an agitation he couldn't really explain. "So you don't find anything at all strange about this."

Connor stared back at him, perplexed. "Okay. You tell me what happened."

Will stood still. *What happened? Angels. Nuns. Monsters.* He shoved the chart away. "I don't know."

Connor watched Will as he circled the cramped lounge. "I don't get what you're looking for . . . what you think—"

"*I don't know.*"

The young doctor sat back on the couch, silent. Will ran a hand through his hair, trying to breathe.

Connor spoke quietly, professionally. "Shouldn't you be thinking about Sydney instead of chasing—whatever you're chasing?"

Will felt his words like a blow.

Connor was immediately contrite, now just a kid again, remembering whom he was talking to. "I'm sorry. I didn't mean—"

Will put a hand over his eyes. "No . . . no. You're right."

Connor stood. "I'm telling you as a doctor. You need to sleep. Go long enough without it and everything gets crazy, and I know what I'm talking about." He hesitated. "Look, you should get Mankau to give you a scrip, but . . ." He stuck a hand in the pocket of his lab coat and pulled out a plastic prescription bottle of pills, extended it to Will. "Use those. If you don't sleep tonight, I want you to have a psych consult. I'm not kidding around here."

Will glanced down at the bottle: Lunestra, prescribed to Connor. He nodded with some chagrin. He put the prescription in his pocket, then reached for O'Keefe's chart on the coffee table. Connor shot him a blistering look and took it from him.

"I'll handle that."

Will started for the door, then stopped, remembering. He turned and looked back at Connor.

"What are those halls behind the elevators? Through the back doors?"

Connor stared at him, and for a moment Will's stomach lurched.

What if he says there aren't any back halls?

"How the hell did you get . . . ?" Then Connor shook his head. "They're service corridors. For transporting surgical patients and—" He broke off, looking uncomfortable.

Will realized what he wasn't saying. "The dead," he said tonelessly.

Connor looked down. "Yeah. The dead halls."

Will turned away.

CHAPTER FOURTEEN

Sleep.

The doors closed, and the elevator started to descend. Will took the pill bottle from his pocket, glanced at the label again. He twisted the top off the bottle, dry swallowed a pill in the faint hope it would kick in by the time he got back to the suite. He leaned against the wall and closed his eyes, knowing he was dangerously close to being dead on his feet . . . not even entirely sure he could make it to the hotel room without collapsing.

Connor was right—what was he chasing?

He was aware he was in some way trying to do his job again, to investigate, to make himself useful, to do something, anything.

Avoidance. That's all this is. Conjuring a reason not to deal with reality.

And what did he really know about medical procedures, anyway?

That officer was dead, his mind argued back.

He opened his eyes as the elevator stopped and the doors opened in front of him onto a long, narrow hall.

Not the lobby of Children's, where he'd been headed. There was no one waiting outside, either. *So where . . . ?*

He stepped forward, glanced down the empty hall. Had he pushed the button by mistake?

He was about to step back into the elevator when there was a flicker of dark in the corner of his eye. He turned from the open doors.

At the far end of the hall, the cop's pregnant wife was walking toward another elevator, flanked by the three nuns Will had seen before. There was something ominous about the sight, almost coercive.

The nuns led the cop's wife into the elevator.

Will's face tightened. He strode down the hall toward the other elevator, slipped inside just as the doors were closing—

There was no one inside. A panel of mirrors reflected him back at himself from all sides, images into infinity.

Will felt a wave of delirium. But this time it took far less time to recover. He moved to the back of the elevator and scanned the mirrored wall, avoiding the crazed look in his own eyes, trying not to think about the incongruence of mirrors in a hospital elevator.

Then he saw it: a button on the side wall, just above the handrail, almost hidden.

He lunged forward, hit the button. The mirrored back wall slid open.

Will stepped out of the back of the elevator, looked around him.

The corridor was similar to the one he'd found last time— brick, dim, with high, arched ceilings. Halls branched out in all directions.

Something behind him creaked in the stillness. He turned—

An orderly was wheeling a gurney with a tiny form, the corpse of a child, wrapped tightly in a sheet like a mummy. Will flinched, stepped backward against the wall. *The dead halls.* Will averted his eyes as the orderly passed by with his unbearable cargo and disappeared around the corner, gurney creaking.

Will exhaled slowly, looked both ways. Down the hall to the left, at the far end, the cop's strawberry-haired wife walked with the three nuns, following a red line on the floor.

Will moved after them.

He walked, with footsteps hollow on the polished tiles, his heartbeat sounding magnified in his own ears as he followed the women through the twisting corridors, staying far enough behind not to be conspicuous. There was not a single other person, doctor, nurse, or patient, in sight.

At the end of a long hall, the nuns and the cop's wife walked through a set of arched double doors.

Will broke into a run after them. He pushed through the doors, feeling a *whoosh* of cool air . . .

. . . and found himself on the mezzanine of a domed, skylit marble rotunda, a part of the hospital complex he had never actually seen before. Several balconies circled the walls below the level Will was on, and a spiral ramp led down floor after floor to the garden level. Gleaming vines spilled from planters on the mezzanine, cascading down several stories. A huge sculptural hand hung suspended in midair; water poured through the fingers into a round fountain far below. The sound of splashing water echoed against the walls.

Will stood for a moment, looking around him . . . uneasy about the feel of the place, though it took him a moment to arrive at even the slightest reason why.

It was lush, almost too lush for its hospital setting, a tropical oasis. And it was enormous: The skylit dome soared countless stories above him, and the ramp seemed to descend below ground level, giving him a feeling of vertigo looking both up and down. How the hell had he managed to miss a building of this size, all these weeks in the hospital?

He crossed to the railing and looked down.

The cop's wife was walking down the spiral ramp, flanked by the three black-cowled figures. There was something in the measured pace—the firm grip that two of the nuns had on her—that made Will suddenly fear for the woman.

Danger. Again, that overwhelming sense of danger.

Will called down, "Wait!" His voice echoed in the rotunda, but none of the women looked up. They disappeared around a curve of the spiral.

Will turned and fairly ran down the spiral ramp after them. Rounding a curve of the ramp, he almost ran into an orderly in white scrubs, straining to push a cart heavily laden with medical equipment up the ramp.

The orderly didn't give Will a glance as he shoved the cart farther up the ramp, the muscles of his bare forearms trembling with the effort.

Will glanced back at the man as he passed and felt another frisson of unease.

Why doesn't he just take an elevator?

Suddenly the orderly buckled . . . slipped to his knees, the overladen cart almost getting away from him.

Will grabbed the cart, halting its precipitous roll, and turned the wheels to the wall of the ramp. He stooped, helped the orderly to his feet. The orderly rose slowly, never looking at Will; he just continued on his way, pushing upward.

Will turned to stare at the man's back as he ascended, muscles straining against his load. The water from the fountain splashed, the echo on the walls around Will now almost deafening. His heart was beating crazily, nearly out of control, a feeling of sheer panic—as if he had been running for miles.

What the hell is going on?

He remembered the cop's wife, gathered himself, and continued down, accelerating, nearly jogging around the last turn of the spiral.

At the bottom of the ramp, he halted, staring out over the floor-level indoor garden. Luxuriant trees and flowers grew from environmental planters laid out in a spiral around the center fountain. Will turned on the tiled path to look around him. There was no sign of the pregnant wife or the nuns; the garden was completely empty. There was a single door in the surrounding curve of wall, closed, with no identifying markings.

Will turned back to the fountain . . . and was startled to see a middle-aged male patient in gold silk pajamas sitting on the edge, staring down into the water. Had he been there all along?

Will moved toward the patient warily. The man was balding, with an unhealthy sagging around his eyes and a sausagelike puffiness to his body. He didn't show any signs of noticing Will's approach. One hand reached toward the water—plump fingers outstretched, but not quite touching it.

"Excuse me, I'm looking for a woman who was walking with three nuns. Do you know which way they went?"

The patient just looked at Will blankly, then went back to staring into the water, fingers vainly extended. The emptiness in his eyes was unnerving.

Streams of water splashed from the giant hand above . . . the ivy curled around the railing, green oversize leaves shining. Will's heart was racing again; he felt a powerful need to flee.

He backed up from the patient—and pushed out through the door in the wall.

To his vast relief, he found himself facing an elevator that stood, doors open, as if waiting. He dodged into the elevator as the door behind him shut on the inner garden.

Will leaned back against the wall of the elevator. His flesh was crawling from the feeling of the rotunda, his heart still pounding a crazed tattoo in his chest. But his mind was surprisingly clear. Fifteen years as a prosecutor made him dead certain that there was something terribly wrong in this hospital.

As his heartbeat slowed, he became aware of the sound of breathing—someone else in the elevator. His eyes flew open . . .

A man stood on the other side of the elevator: rather tall, though not quite Will's height, longish arms and legs, the crew cut of a soldier. Will stared at him in puzzlement and was rocked by a wave of déjà vu; more than that, a sickening, vertiginous feeling. . . . But he hadn't a clue where he knew the man from.

Then he recognized him.

It was the limbless man he'd seen in the elevator and on crutches in the hall.

Only now he had arms and legs.

He was standing in the corner, where his wheelchair had been. Whole, dressed, the same pale, intense eyes—not burning with rage now, but veiled, wary, almost hostile as he met Will's gaze.

Will stared, numb with disbelief. The elevator doors opened, and the man walked out.

The doors shut before Will could speak. The elevator continued its ascent.

CHAPTER FIFTEEN

For the first time since Sydney was admitted to the hospital, Will walked into the cocktail lounge of the galleria.

The bar was larger than he'd expected, comfortingly dark and inviting, with a plush maroon carpet and gold-framed mirrors reflecting gleaming dark wood and bottles.

Will stood, looking into the darkness. Aside from good wine at dinner, he wasn't a drinker. He'd been too conscious of the ravages of his father's three-martini lunches and his mother's genteel alcoholism to seek out that kind of crippling dependence for himself. And from the moment he'd met her, Joanna had been his drug of choice.

He could well imagine that upon receiving the death sentence of a loved one, as doled out daily by the hospital, many relatives dove straight into a bottle and never came out. He'd experienced the opposite effect: He hadn't once thought of alcohol since he'd come to Briarwood. It would put too great a strain on a reality already warped beyond recognition.

But after the incident in the rotunda, the limbless man impossibly restored to wholeness, it was clear that he'd passed some point of no return. No amount of drinking could compare with the last surreal hour.

Will walked to the bar. A bartender in his forties came over to him, and Will ordered briefly, "Scotch." No need to specify; the bartender had registered the cut of Will's suit, most probably had recognized him from the moment he'd walked in. He reached behind him for the Glenlivet, put a glass in front of Will, and poured.

Will sat on a stool and was mildly surprised at the relief he felt in his legs. He'd been on his feet for who knew how many hours. The thought of legs brought his mind back to the man in the elevator, no longer legless. He grabbed the Scotch and drank.

The first sip was like slipping into a warm bath. Another swallow and his nerves stopped screaming for sleep. For the first time in months, he felt above the pain.

He drank again and didn't protest when the bartender stepped in discreetly to refill his glass. *What hasn't he seen? Working in a hospital bar must be its own special window into hell. . . .*

Will took another deep drink, and the thoughts retreated. Now riding a blissful feeling of calm, he gazed ahead of him at the surprisingly elegant curved antique mirror above the bar, past the bottles into the silvery glass.

At the corner of the bar, two doctors sat in their scarlet scrubs. They glanced at him furtively as they talked, then up at the TV hung above the bar, on low.

Will looked up at the screen—and saw his own face. A newscaster mouthed the context; Will read along with the crawl: "Campaign headquarters refused to confirm rumors today that Will Sullivan, once considered a shoo-in for the primary nomination, is dropping out of the governor's race . . ."

The screen seemed a snapshot of another life.

The camera cut to moving footage of Will, and he regarded himself as from a great distance. The man he observed on the screen was confident, charismatic, a born leader. Will fought an urge to laugh out loud. *I'd vote for him. No question.*

He clutched the glass in his hand to keep from hurling it into the screen. If indeed he was going crazy, it was all the better that he'd withdrawn. There was no question of holding any kind of office in this condition.

Will looked down into his glass and took another swallow . . . then looked up at his reflection in the bar mirror, the ravaged expression in his eyes. His body shuddered with an involuntary spasm. And in his mind, he was screaming for—help—release—

Help.

In the mirror, he saw a man walk into the lounge, moving toward the bar. Tall, elegant, unmistakable.

Will turned toward the door—

No one was there.

Will sat frozen, not breathing. Then he heard a sigh beside him. He whipped around. Salk stood on the other side of him at the bar, loosening his tie, an easy, sensual gesture.

Will blinked. After a moment, time began again. He nodded, and at the cue, Salk smiled, that rueful smile, and moved down to join him.

He glanced at Will's drink and said simply, "Hard night."

Will looked at him. His own voice seemed hoarse to him as he spoke. "You said you'd worked at this hospital a long time."

Salk shook his head. "Ages."

Will put his hands on the edge of the bar to steady them. Salk watched him.

"Why? What is it?"

Will glanced away, took a breath. "There's something wrong here."

The bartender had moved forward and was hovering discreetly, more in front of Will than Salk. Will nodded to his own glass. "Another—and . . ." He looked at Salk.

Salk nodded his thanks. "A Harp..." Then his eyes met Will's. "Unless I'm still on duty."

Will hesitated. Salk stood, gestured toward a booth in the back of the bar.

"Step into my office."

They both had removed their ties. Several glasses were on the table in front of them. Mirrors around them reflected multiple images of them.

Will was talking, could not stop talking, the weirdnesses of the last few days spilling out of him randomly, as if he really were drunk, or insane. Gary in Sydney's room. The three nuns with the cop's wife. The orderly in the rotunda. The increasingly oppressive feeling of conspiracy, malevolence, of something dangerous and perverse in the hospital.

"I could accept that I only *thought* I saw a statue move, and a monstrous nun. But just now—a quadruple amputee suddenly has arms and legs?" Will clenched his glass so hard in his fist, he could have broken it. "Impossible... but I *saw* it."

Throughout Salk had sat quietly, merely listening. Will finally sat back, light-headed from the release of finally saying it all. It was madness, but it was out.

Now Salk leaned forward with careful urgency. "Mr. Sullivan, this hospital is such an enclosed world, so cut off from reality ... so many people around you in pain. Add to that your own extremely traumatic situation. It's not at all uncommon to imagine things—even hallucinate—under the prolonged stress of a hospital stay. Not unlike what happens with sleep deprivation, which you obviously have a pretty good case of by now...."

Sleep deprivation. First Connor, now Salk ...

Will stared at him. "You're saying—you've seen this before? Like *this*?"

Salk half smiled. "There's even a name for it. The hospital calls it 'sundowning.'"

A name for it? An actual diagnosis?

Will sat back, nearly sick with relief. "God. I really thought—I was going—"

"Crazy. I know. They started your daughter on a new treatment today, didn't they?"

Will remembered Salk in the doorway, the sensation of being watched. Salk spread his hands. "That's precisely the kind of stress that would bring on sundowning."

Will ran through it in his head. The sleeplessness, the paranoia, the sense of being off balance; the ominousness around him, the strange sights. Certainly he'd seen his share of post-traumatic stress disorder, so in theory this was nothing new. But . . .

Salk must have seen the doubt in his face. "Is there something else?" Will hesitated. Salk prompted him gently. "You've been asking questions about Officer O'Keefe."

Will thought with sudden paranoia that Connor must have mentioned the business with the cop's chart, perhaps had even sent Salk after Will, for the consult he'd mentioned.

All right, so he was on a dangerous edge. But that cop was no hallucination.

"You *saw* him," Will said, his jaw tight. "He was dying, nearly dead. And then the next day—the next moment, almost—he was walking around as if nothing had happened."

"And why is that significant to you?"

Will stared at the other man, incredulous. "How is it *not* significant? The man was dead. It's a miracle. I thought . . ." He paused. Salk merely looked at him, waiting. "I don't know what I thought."

Will swallowed the dregs of his drink. He was beginning to feel drunk.

Salk spoke gently. "Yes, you do. You just said it. You thought it was a miracle." He found Will's eyes, held them. "Mr. Sullivan, you're looking for a miracle."

Will felt a warm flush all through him. He knew the feeling from court.

It was truth.

At some point they left the bar—perhaps after last call, or perhaps there was no last call in that particular way station. Now the two men walked on a glass bridge, high above the lights of the vast hospital complex.

"Of course you want a miracle. What parent wouldn't?" Salk's calm voice was almost hypnotic.

"Of course," Will echoed dully. He'd thought he was evading the reality of Sydney's impending death by manufacturing a mystery. But he'd been looking for something he hadn't even let himself admit. He'd been looking to save her.

"And of course you're a Catholic," Salk said. It was not a question.

Will stiffened, started to shake his head. It had been the night Joanna had told him of her childhood that he'd felt any vestige of faith slip away from him forever.

"Very long lapsed," Will said, his voice cold.

"Who isn't?" Salk agreed. "And yet . . . you were raised with the possibility of miracles. It is a pillar of our religion. Of every religion. In times of stress we revert to our core beliefs." He shrugged. "It's only human."

Will spoke flatly. "So I've been making all of this up. It's entirely my imagination."

"Not at all."

Will turned toward the dark man, not understanding.

Salk spread his hands. "Officer O'Keefe's recovery *is* a miracle. How could any reasonable person say it wasn't? The man was dead."

Will stared. He was relieved beyond measure at the confirmation—but also had the prickly sensation of moving into uncharted territory. Around them, the bridge seemed oddly to have lengthened; they'd been walking for minutes and were not even halfway across.

"Yes . . . exactly."

He looked at Salk, now confounded. Salk half smiled, as if reading his thoughts. "Of course I'm not indifferent to it. It's my life's work. And the impossible *does* happen . . ." The counselor paused, choosing his next words carefully. "But not every day, and not for everyone. For some people, and not for others." He paused again, and his voice dropped infinitesimally. "Do you think that's random?"

Will frowned. *Not random?* "What are you saying?" he said stiffly.

"Take the very example you've been pursuing. Officer O'Keefe has an enormous will—an intense desire to survive. A pregnant wife. An unborn child he was determined to live for. And I might add, his wife's desire for him to live was even stronger. Perhaps that desire made the difference."

Will looked out through the glass of the bridge, into the dark sky. He was aware of a struggle going on inside him, a violent surge of emotion, like a warning . . .

Danger?

But why? It was nothing he hadn't heard before—that New Age line. Positive attitude, creative visualization, "alternative" healing . . .

Still, his whole body was clenched, his face hard, as he answered, "I think it's a dangerous thing you're saying."

Salk watched him with a faint smile. "Dangerous—to hope?"

Will's voice was like ice. "Dangerous to think we have that kind of control over things that are out of our hands."

He knew he sounded defensive. As if they hadn't done everything in their power for Sydney. The best hospital, the best doctors, no expense spared. As if they somehow hadn't loved her enough—

"No one would ever say you hadn't loved your daughter enough."

Will was not sure if he had spoken aloud, or if Salk had read his thoughts, or if he had only imagined Salk speaking. He looked out through the wide dark glass of the bridge, down into the frozen garden, at the bench where he'd found Joanna the night before, the willow tree . . . the angel . . .

Salk's hand was suddenly under his arm, grasping it, and Will realized that without the support he would have collapsed. Salk smoothly guided him to a bench on the long bridge and hovered as Will folded himself onto the seat, caught his breath. He suddenly remembered the sleeping pill he'd taken before the—how many glasses of Scotch?

"I haven't eaten," he said when the floor had stopped spinning.

Salk stood above him. "It's more than that. Something about this conversation is causing you profound disturbance. What is it?"

Will was shocked by the counselor's grasp of the situation. Without meaning to, he spoke. "My wife . . ." He had to stop, gasp in.

Salk spoke beside him, encouraging. "Your wife . . . ?"

Will felt a wave of guilt so strong, he was nearly sick. "God . . . my daughter's *dying,* and all I can think about is my wife. . . ."

Salk interjected quickly, reassuring. "But of course. They're both everything to you."

The guilt washed through Will again. "You don't know."

Salk studied him with a faint, puzzled frown. Then he laughed, startling Will. "Mr. Sullivan, it's no crime to be in love with your wife."

The words were a comfort almost like pain. Suddenly Will found his terror pouring out in a rush.

"This is killing her. She's not rational. I can't reach her at all." He shot to his feet, pacing on the bridge. "She thinks she can do something. About Sydney. She . . ." His voice dropped to barely a whisper. "She tells Sydney she's not going to die."

The words hung in the silence. The night seemed to deepen outside the surrounding glass.

Then Salk lifted his hands and spoke serenely. "But what if she's right?"

The words were like a physical blow to Will, like a wave crashing over his head, tumbling him in the surf. Salk's voice continued inexorably. "What if she *can* save her? What if you both can?"

Will felt himself staring as if Salk were crazed, or he were, or both. Salk was unfazed. "Children respond strongly to their parents' beliefs. It's not unlike hypnosis. Your wife may well be giving your daughter the will she needs to turn her own illness around."

With anyone else, maybe, maybe! Will wanted to shout. "No. No. It's not strength."

"What is it, then?"

"It's guilt."

Salk raised an eyebrow but said nothing. Will forced out the words. "She thinks—Sydney's cancer is her fault."

"Her fault? How?"

Will leaned on the windowsill railing of the bridge. The night sky was black above him. "Her father . . ." He stopped, shaking, unable to go further, to betray Joanna to a stranger.

But there was no need. Salk was quiet for a moment, then said simply, "There was abuse."

Will's face was taut. He clenched his hands to keep them from trembling. "If he weren't dead, I would kill him. I would make it last a long time."

He stared out into the dark, over the eerie floating lights of the hospital complex, fighting to keep the rage down. Shadows he didn't want to see crowded in on him: visions of a child cowering in the dark, then weeping in pain and shame—things he would gladly have died to spare her from.

As he forced himself away from the thoughts back to the present, he was aware of Salk watching him gravely.

"So your wife believes that because she is damaged—your daughter became damaged."

Will turned to the other man, startled by Salk's uncanny precision.

Salk's eyes were infinitely complex. "Of course. I've seen it before. That kind of wound lasts a lifetime."

Will's voice was hoarse. "I'm afraid this is killing her."

"Perhaps Joanna is stronger than you think."

Will looked at the other man sharply, but his expression was as neutral as his tone had been. Will's voice dropped.

"Sydney is everything to her. From the moment she was born . . . they're like the same person." Then he found himself confiding something he'd never admitted to anyone—the guilt that haunted him. "Sometimes I feel left out."

He had never before said it to a living soul . . . had never even really let himself think it. Salk's gaze on him intensified, a quickening of interest.

"And since Sydney's illness?"

"We don't talk about it. She won't." His voice lowered with his secret shame. "She thinks I've given up."

"Mr. Sullivan . . ." Salk hesitated. "Will. You are a practical man. Reasonable. Logical. You have been distancing yourself from your daughter because you believe she is going to die." Will's heart plummeted at the accusation. Salk added reassuringly, "That's natural; it's simple survival." The dark man's gaze did not judge but did not quite absolve him, either. "But your wife does not care about logic, or reason, or reality. If someone told her she could save Sydney by going out under the full moon and sacrificing an animal, by spilling her own blood, she would do it—without question."

Will stared at him, disturbed. "But that's . . ." He couldn't bring himself to say the word.

"Madness?" Salk said. "Is it?" Blood was pounding in Will's head. He was riveted to Salk's words.

"Crisis is a turning point. It is in its way an altered state—it brings out the true soul. There is enormous power in that point of concentration. Human beings have always had the ability to change their circumstances, to change what we think of as reality—by thought . . . by faith . . . by will."

Will could not move, could not look away. "You really believe that?"

Salk was unsmiling. "I know it." His slate eyes were dark, almost black. "I've been at this a long time. I have seen astonishing things happen. Diseases reverse themselves. Terminal patients recover. Human beings are infinitely resourceful, and the world is full of mystery."

Time seemed to have stopped; there was nothing on earth besides the two of them. Will felt drunk with the intensity.

"You're honestly talking about—miracles."

"But why not a miracle? It happens. You've seen it. That's why you're here." Salk looked at Will, spoke softly. "Why not take it?"

The need to believe was so strong, Will felt himself teetering on a great balance, with a desire to let go, to fall . . .

He spoke through a dry mouth, his voice sounding to him a million miles away. "What would it take?" He was not entirely sure what he was asking.

He could see only Salk's eyes boring into his. "What are you willing to do?"

The sudden sensation of nausea was almost paralyzing. Will could barely speak. "I don't understand what you're saying."

Salk smiled, and his teeth were whiter and stranger than Will could process.

"Don't you? I think your wife would understand."

The words broke Will's trance. He lunged forward, grabbing for Salk's lapels, an atavistic reaction that he could not properly have explained to himself. He had the impression that Salk had sidestepped him, yet he was holding Salk's coat and staring into the other man's face.

Something flashed in Salk's eyes, like flame. But perhaps it was only a trick of light. The dizzying sickness Will felt was so strong, it seemed to him the whole bridge was spinning.

Then the other man stepped back from Will, spoke ruefully, regretfully. "I see I haven't convinced you." He inclined his head. "Perhaps . . . one day."

Outside the wall of windows, there was a glimmer of light in the sky above the hospital complex. Salk half smiled, with none of the strangeness Will had imagined. "Come by my office anytime you want to talk. I'm always open." He indicated a red line on the floor. "Leads right to my door."

He turned and walked the length of the bridge, tall and elegant. Will stood with his back against the glass, his muscles trembling in exhaustion as he watched Salk go.

CHAPTER SIXTEEN

It had seemed only an hour that he'd been with Salk, but dawn was breaking when Will finally arrived back at Sydney's room.

He felt as if he were just awakening from a disturbing dream, part of which he couldn't recall. His whole body was as tight and knotted as if he'd been engaged in a life-and-death battle.

Yet nothing had happened. He barely remembered the conversation, much less why he would have had such a physical reaction to it.

No more drinking, he thought grimly. *And no pills. Not on top of everything else.*

He stepped through the door of Sydney's room, and his heart wrenched as he saw his baby sleeping. *So small . . .*

In his head, Salk's voice suddenly spoke, a mocking snippet of memory from the night before. *You have been distancing yourself from your daughter because you believe she is going to die.*

Will closed his eyes, fighting the words in his head. Sydney stirred on the bed as he stepped into the room. He glanced reflexively into the corner. Joanna's cot was empty; there was no sign of her.

Again.

He resisted the urge to leave the room immediately to find her. Instead, he moved past the cot to the bed and sat beside Sydney, cupping her head in his hand as the day lightened outside the window.

He whispered, not knowing if she heard, "I love you, baby."

Sydney's eyes opened instantly, startling him. She looked up at him steadily, a clear gray gaze. "I know," she said with surreal calm.

Will's smile trembled. His eyes wandered to the empty cot.

"Where's Mommy, sweetie?"

"She went with the man."

Will's blood turned to ice. *No. Not possible,* he thought, even as rage rose in him.

"Who?"

Sydney didn't speak.

"Dr. Connor? Dr. Mankau?"

Sydney sighed, an exaggerated breath. "No. The *man.*"

Will stepped out of Sydney's room, looked down one side of the corridor, then the other. Then, knowing, he crossed the hall to the wide window and looked down into the snowy garden.

Joanna sat on the marble bench under the frozen willow tree, with the angel statue hovering across the path. She was in intent conversation with a man beside her, in the shadows of the tree.

Salk.

Joanna was very close to him, their bodies almost touching. She listened to Salk, nodding, completely open to him—no sign of the wariness she showed with all people. Salk's intense, innate sensuality was something Will had noted only in passing. Now even from this distance, his powerful charisma was blindingly present—the way he leaned in to Joanna, long fingers lightly touching her wrist, her thigh.

A white hot fury swept through Will. Salk's words blazed in his skull.

I think your wife would understand.

Salk looked up, as if feeling Will's rage. His eyes went straight to the windows . . . to meet Will's gaze.

Without taking his eyes from Will, he leaned in close beside Joanna's ear and spoke into her hair.

Will turned from the window and sprinted down the corridor.

He shoved out through the glass doors into the snowy garden and strode through the children's section, then on the icy gravel path through the row of statues. The stone figures towered around him. Will stared straight ahead, refusing to look. He hurtled past the last statue in the aisle—

And halted.

Joanna sat alone on the bench beneath the willow tree. Her gaze was fixed on the sculpted angel on the other side of the path. No one else was there.

Will walked slowly toward her, looking around the empty garden. Just the statues . . . and the frozen trees. Joanna turned slightly as he halted beside her. She looked at him as if she'd been asleep.

Will's voice was taut, barely under control. "Where is he?"

Joanna gazed at him blankly. "Who?"

Will's breath stopped. He looked at her for a silent beat. She stood from the bench, defensive. Will's voice was tight; he could barely keep from shouting.

"What was he saying to you? How long have you been seeing him?"

Something flashed in her dark blue eyes. "*Seeing* him? We were only talking—"

"Joanna, you just lied about him—"

"Because I knew you wouldn't understand," she cried out. Will stood, waiting. She sank onto the bench. "I'm here all the time.

Sometimes I think . . . I'm going crazy. He asked about Sydney. He—talks to me. That's all."

Her voice dropped, shaky, almost in tears. "He makes me hope." *He talks to me,* she'd said. Not once—plural.

Fury swept through him as he realized how it must have happened. The same chance meetings. Salk's oh-so-subtle lines about finding power in crisis.

How instantly he must have hooked her.

"Oh Christ, Joanna. He's up to something." Will could hear the ugliness in his own voice, but the ruse outraged him. His own gullibility outraged him. He was a prosecutor; he should have *known*. "He's been talking to me all along as if he's never spoken to you. Why?"

He laughed, a hollow sound. He was aware of his own helpless jealousy—that this man had been comforting his wife, false comfort though it was. That Salk had been able to reach Joanna in a way Will hadn't.

"And you've been taking him in to see Sydney. This . . . charlatan, pretending he has some kind of say about whether she lives or dies . . . and making you believe it."

Joanna shot to her feet. "Yes. Because you've given up. You'll let her die. You're just waiting for her to die." Her eyes were blazing. Will stepped back, stricken. He could barely speak.

"You can't—think that—"

Joanna's voice was raw. "I don't care what you or the doctors or anyone else thinks. She's not going to die because *I won't let her*."

There was in her eyes what Will had been fearing all along. Madness. The precipice.

She bolted from him, down the path toward the hospital.

Will stood alone with the angel, unable to move.

CHAPTER SEVENTEEN

W ill slammed into the counseling center, located on a fourth-floor corner of yet another hospital building, the Fordham Clinic. The blond-wood reception desk had the lines of a temple; the decor was Zen-like abstract art and planters of bamboo, a designer's attempt to create the spare feel of a Japanese garden.

The receptionist started to smile up at Will, but the smile died when she saw his set face.

"May I help you?"

"I want to talk to Salk."

The receptionist tilted her head, frowning. "I'm sorry, who?"

"Salk. One of your counselors." The fury was harsh in Will's voice.

The receptionist glanced nervously toward a row of doors along the wall.

"One moment, please, Mr. . . . ?"

"Sullivan," Will said shortly, and waited for it to sink in. The receptionist's eyes widened.

"Oh! Of course, Mr. Sullivan. If you'll just . . ." She stood, flustered, and went into one of the doors along the wall. Will pressed both hands on the edge of the counter, trying to breathe through his rage.

Through a door in the opposite wall, a meditation class was in session; a female instructor in soft white clothing walked between patients seated in the lotus position on yoga mats, speaking in a low, carrying voice. "Focus on the light. The light surrounds you . . . breathe in the light . . ."

Will turned away as a door opened behind him. A woman stepped out, in her fifties, a little too determinedly colorful in her vaguely Indian caftan. She crossed to Will, warmly extending both hands. "Mr. Sullivan, I'm Martha Lillian, the director of our counseling center. I'm so glad you dropped by—"

Will interrupted, barely maintaining civility. "Excuse me. I came to speak with Salk."

Lillian looked mystified. Will knew what she was going to say an instant before she spoke: "We don't have a counselor named Salk."

Will stared back at her. All he could hear was the pounding of his pulse in his own head.

He paced the floor of the lavishly paneled hospital office. Oil portraits of the hospital founders gazed down from the walls. Will hadn't been inside since Harcourt, the chief administrator, had invited him in and offered him his choice of the residential suites, anything the hospital could do for them, anything at all.

Now Harcourt sat behind a wide oak desk, looking more like a banker than a hospital director. The officious hospital attorney watched Will nervously, smelling a lawsuit.

Yeah, you better be worried, Will lasered at him silently.

Martha Lillian sat on the edge of a silk-upholstered sofa, caftan tucked around her knees, reporting to the men. "After Mr. Sullivan brought this to my attention, I checked all our hospitals.

There are no counselors named Salk, or fitting Mr. Sullivan's description."

The lawyer spoke up. "Did he actually *say* he worked at the hospital?"

Will gave him a look that would melt steel. But inside, he paused. *Had* Salk ever directly said he was on staff? Will's mind scanned through what he remembered. He'd seen Salk with so many patients. Could he have simply *assumed* Salk worked there?

Nevertheless, there had been clear, deliberate deception.

Will turned to the chief administrator. "If you're telling me you have no Salk on staff, then you have some con artist walking around your hospital misrepresenting himself to your patients, exploiting people's pain. He's been passing himself off as a counselor to my wife—"

Will broke off at his own words as a thought occurred to him that he'd never considered before. This was about *him*. Salk was specifically targeting him and Joanna.

He sat, a multitude of possibilities swirling in his mind. The administrator and the lawyer exchanged a cautious glance.

"Mr. Sullivan, we'll do everything in our power to find out—"

Will cut the administrator off. "No."

The other two men waited, without moving, while Will looked out the window for a moment, thinking. He finally said in a voice that precluded argument:

"I'll take care of it myself."

He walked grimly out the white-columned entrance of the administration building into the icy wind outside and stopped on the stairs, breathing in.

His sleepless mind clicked into prosecutor mode and began to process. Salk was a fake—a very elegant, expensive fake. He'd been working on Will and Joanna at the same time.

Joanna, who in her desperation was open to almost anything. . . .

Dark suspicions crept into Will's mind, thoughts he didn't want to allow.

The question was, how far had it gone? What was Salk promising her? And what precisely did he want?

Will turned, looked back at the massive hospital complex, and suddenly he knew. He pulled out his cell phone, punched a speed-dial number.

Jerry's voice answered with a manic cheeriness that indicated he was in full fund-raising mode. Will could picture him in the store-front office they'd set up at the beginning of the campaign, surrounded by SULLIVAN NOW posters, working two phones at a time.

"Jerry. Are you alone?"

Jerry's voice bubbled over. "*Will*. My God, this is fantastic. The interview you gave to Dawes—why didn't you tell me?"

Will had no idea what he was talking about. Dawes? The *Boston Sentinel* reporter?

"Front page, three columns. AP's picked it up. . . ."

Will spotted a newspaper discarded on one of the marble benches lining the portico. He moved over to pick up the front section, looked down at the headline:

"I HAVE HOPE," SAYS SULLIVAN

"Unbelievable," Jerry enthused. "Money is pouring in. E-mails, cards, flowers—people are just going crazy over this—"

Will cut through his words. "I need you to get a PI on something. Use Hank."

Jerry was instantly alert, focused. "What? Why?"

"There's a man who calls himself Salk—S like in Sam, A, L, K—who's been hanging around the hospital passing himself off as a counselor. He comes on like . . ." He paused, groping for words. "A faith healer. Not a fire-and-brimstone type, one of those quantum universe, power-of-will, mind-over-matter guys.

Charismatic." He felt another surge of rage at the thought of what Salk might have promised. "I've seen him talking to other patients, but he's focused on us, Jerry. Joanna and me. He's been approaching Joanna behind my back."

There was a brief pause on the other end, then Jerry's voice—neutral, calculating. "That could be bad."

Will matched Jerry's tone. "I'm wondering . . . who would go to this much trouble to set me up? Who would hire someone like that to fuck with me?"

He could feel the thickness of the silence—Jerry, immediately thinking through the consequences. It was not, after all, a completely surprising scenario. Will was kicking himself for not having caught on before. He himself had always lived a practically perfect public life. He'd understood since before he was even a teenager that a career in politics meant every aspect of his life was subject to microscopic scrutiny. There were no skeletons in his closets. Joanna's comparatively mysterious past had always been of intense interest to Will's opponents—men sensing a possible point of vulnerability, even scandal. At the beginning of the campaign, Jerry had told Will bluntly that his opponents would not hesitate to go after Joanna if they couldn't find anything on him.

"You think it's Benullo's people, digging for some kind of dirt," Jerry said flatly.

"Exactly," Will said. "I don't care what it takes. Find out what his game is and put a stop to it. Quietly."

Jerry took control. "I'm calling Hank now. We're going to nail these bastards." Will was aware that his campaign manager's response was fueled partly by the fact that it suddenly looked very much as though Will were back in the game. For the moment, Will wasn't going to correct him on that.

He punched off the phone. Jerry would deal with Salk.

He had more important work to do.

CHAPTER EIGHTEEN

Cold shadows lengthened across the garden paths in the dusk. The garden was ice; colorless, lifeless.

Joanna was a specter in her black coat. She sat on the bench under the willow tree, very still. Her eyes were fixed on the angel statue . . . she twisted a pair of black velvet gloves in her hands. There was an active quality in her waiting, her body poised, alert . . .

Out of nowhere, hands reached down to touch her hair, her face. She looked up . . .

At Will.

He knelt in front of her, slid his arms around her waist, laying his head in her lap, breathing in her fragrance of cold and perfume.

Her fingers moved in his hair, slightly, abstractedly. His heart beat faster at the simple, familiar gesture. It had been so long since they'd touched.

She bent her head down to his, smelling his hair.

"So clean."

There was something different in her voice: different in that it was her own voice, her real voice, absent the pain and distance that had overtaken her.

Will lifted his head to look at her. She touched his face, but her dark blue eyes were somewhere else, remembering.

"That's what I thought when I first saw you. How can anyone be so clean? It was like light . . . inside you. I wanted . . . to be inside that light. . . ."

Will shook his head, his voice thick. "You're the one—"

His voice broke; he was on the edge of tears. The twilight deepened around them, but he raced ahead urgently, afraid to lose the moment, the connection. "I'm trying to believe, Joanna."

Her fingers found his lips. "Shh. It's all right. I—" She touched his face, bent to kiss his mouth. "I love you. Never . . . doubt that."

Now Will's eyes did fill with tears. He groped for her hands. "I never have."

He gathered her in his arms, holding her fiercely. Her body melded into his, and he felt her there, warm, and real. For the first time, he felt that no matter what happened, they would face it— together.

Then both their beepers went off at once, splitting the silence of the garden with an electronic scream.

CHAPTER NINETEEN

Everything that followed was a nightmare, an altered state—a flurry of scenes, impressions, at once sped up and in slow motion.

Will and Joanna ran through the twisting hospital corridors, past the long wards of Alzheimer's patients, past the struggling recoverees in physical therapy, startling visitors and drawing knowing looks of sympathy and dread from nurses and parents. The corridors seemed to elongate in front of and behind them, crazily, as in a dream.

Will burst through the doors of the surgical waiting room, Joanna right behind.

The lavender waiting room was overflowing with members of what seemed to be the same extended Italian family. Through his panic, Will recognized the Italian girl he'd seen praying over her young brother's bed, now luminous with terror. There was a tiny, dark grandmother with a rosary, a muscular father looking edgy and helpless, various aunts and uncles and children—some weeping openly, others whispering in Italian.

Will strode through the gathered family to the nurse at the desk. "Will and Joanna Sullivan—we were paged—"

The desk nurse dropped her eyes evasively. "I'll tell the doctors you're here." She reached for the phone.

Will leaned on the desk. "What's happening?" Before the nurse could answer, the double doors opened behind the desk. Dr. Connor stepped out in scarlet scrubs. Will crossed to him in a shot.

"Connor. What's happening?"

The young doctor put both hands on Will's arms—a surprisingly strong grip—and spoke gently, quietly. "The tumor's blocked her kidneys. She's gone into renal failure."

Will's legs turned to water underneath him. "God. No . . ."

Connor was speaking. Will could barely hear him over the pounding of his own heart. "They're prepping her for surgery now. I'll be back as soon as we know more."

Will swallowed. Connor tightened his grip on Will's arms. "Mankau's the best," he said softly. The faint comfort was all the more terrifying.

Connor disappeared back through the double doors. Will turned blindly toward Joanna and felt his heart contract.

She was gone.

He scanned the milling Italian family. A door on the side wall was closing slowly.

Will strode across the waiting room and grabbed the door handle, pushing through the side door.

He was in a long, empty, rectangular hall, a windowless dark purple, with eight sets of elevator doors on all sides of him. One of the sets of doors was just sliding shut.

Will catapulted across the hall and muscled through the doors.

The elevator was empty. Mirrors reflected him back at himself from all sides.

Will advanced on the back wall, searching for the concealed

button to open the back doors. Nothing. His hands scrabbled along the length of the mirrored wall. He was wild with fear, not knowing what he feared, only that he had to find Joanna before it was too late—and not knowing what too late meant.

He lunged at the front wall and banged his fist against the control panel, then the walls, punching and kicking at his own crazed reflection, as if sheer violence could bring her back.

There was a mechanical whirring behind him. Will swung around.

The mirrored back walls of the elevator slid open.

He stepped out of the back of the elevator and turned around him, searching the halls. They stretched out, endless and empty. No sign of Joanna.

Out of nowhere, three orderlies flew past, pushing a gurney bearing a patient. The patient was screaming horribly; blood dripped from the gurney, trailing along the floor.

Will jumped back as they ran past him, almost knocking him down. They disappeared around a corner. The screams stopped abruptly.

The silence was worse than the screaming.

Will pulled himself together and looked around him, saw the directional lines painted on the floor—yellow, blue, green, red, like a road map, or veins. Will stared down at the red line and heard Salk's voice:

Come by my office anytime you want to talk. . . . Leads right to my door.

There is no office! Will shouted in his head. But Joanna would have believed it.

He started off after the line, after her.

The red band twisted through corridors, around corners . . . it seemed to Will almost to pulse, itself, in time with his own blood. The pulse quickened as he strode faster, following the red line through the maze, past open doors of rooms. Sounds drifted from the doors, following Will, building, overlapping: crying, soft moans of pain . . . the whisper of prayers.

Will dashed around another corner—and froze. Three black shapes hovered in the intersection of the opposite corridor. The nuns paused for a moment, looking down the long hall at Will, their faces in shadows under their headpieces. Then they turned and walked along the red line, disappearing around a corner, black habits rippling like water.

Will strode down the red line after the nuns. He turned the corner into a dimly lit hall—and almost gasped aloud.

The nuns were impossibly far down the hall.

Will broke into a run. The nuns glided through double doors at the end of the passage.

Will pounded down the hall, pushed through the doors—and stopped.

He was in the skylit rotunda with the spiral ramp, the wrap-around balconies, the enormous sculptural hand with water pouring from its fingers.

Will lunged toward the railing. The nuns had disappeared. But through the tangle of gleaming vines, he could see the same orderly on the ramp far below, straining to push his cart upward.

Will started grimly down the ramp. At his feet, the red line wound sinuously down the curves of the spiral.

As Will rounded the spiral toward the orderly, the orderly buckled . . . fell to his knees, the overladen cart slipping away from him.

Will grabbed the cart, braced it with his leg, and helped the orderly to his feet. The orderly never looked at Will, just continued on his way, pushing upward.

Will felt for the handrail, gripped it. The wave of déjà vu was almost suffocating, but he bolted forward, descending. *Find Joanna.*

He rounded the last curve of the spiral and looked out across the garden. The patient in gold silk pajamas still sat on the edge of the fountain, arm outstretched toward the water.

Will cleared his throat to speak, and the patient jerked his head up. His face was horrifically shrunken . . . his tongue blackened, lolling . . .

Will turned wildly in the spiral garden, looking for the red line. It stopped at the single door in the wall. He ran to the door and pushed through it—

He was facing the wooden doors of the hospital chapel on the right and the glass wall of the outdoor garden on the left. The green-tinted hall was empty, still. . . .

Will looked toward the doors of the chapel.

Then the doors to the garden burst open, and Teresa Marinaro stumbled out, like a puppet staggering onto a stage.

Will caught her, steadying her. Her face was unfocused, terrified.

"*Signora . . . aiutami,*" she whispered.

Will's hands tightened on her arms. "What is it? What's wrong?" he asked, but she pulled away from him and bolted around a corner, her running footsteps echoing, then fading into silence.

Will looked out the glass wall into the shadowy garden. He moved grimly toward the doors and pushed through them into the icy night.

The cold hit him like a shock. Snow swirled in flurries around the elf and gnome statues in the children's section of the garden. Clouds scudded over the full moon; in the preternatural light, everything seemed alive. The wind whispered through bare branches like a chant.

Will ran forward, crunching snow like glass under his feet, breathing plumes of frost . . . down the gravel path between the skeletal trees . . . through the line of looming statues, ice frozen and glistening on their faces. The last statue reached for him with outstretched arms. Condensation dripped from its eyes, like tears.

Will stopped, staring toward the willow tree.

The bench under the tree was empty. The garden was perfectly silent and white.

Will breathed in frozen air, walked slowly to the bench . . . reached down, and picked something up from the snow.

A black velvet glove.

CHAPTER TWENTY

A loudspeaker was blaring in his ears. Will stood in the doorway of the lavender surgical waiting room. But had he walked back from the garden? Or had he ever left at all? He looked around him in total disorientation. Joanna was nowhere in sight.

The Italian family was gathered around in a flurry of panic, staring at the speaker in the ceiling, Teresa standing just as before, pale in the center of her cluster of relatives. Will looked toward her, hoping for some kind of corroboration, but she would not meet his eyes.

"Dr. Murphy, Dr. Ignacio, Dr. Rosen. Code, OR Five."

Will had no time to think, no time to wonder if he'd dreamed everything. A surgeon pushed through the double doors. The room fell silent—Teresa and her family on one side of the room, Will alone on the other.

No one breathed.

The doctor looked toward Will, removed his mask. He was Latin, hawk-nosed and copper-skinned.

Not Mankau.

The surgeon walked slowly to the Italian family, stopped in front of the father.

The father was trembling, his voice thick. "My . . . son . . ."

The surgeon spoke. "I'm sorry. We did everything we could."

Silence.

And then Teresa was engulfed by her family, and sobbing broke out in waves. Will turned away, ashen.

The double doors opened again. Dr. Mankau stood in the doorway in his scarlet operating gown, without his mask. It took him a lifetime to cross the floor . . . to stop in front of Will. Will was paralyzed with dread.

The Indian spoke in his musical accent. "We have taken the pressure off her kidneys. We are getting some circulation. She is stabilized. We are very lucky."

Will breathed out in wrenching relief. The sobbing and praying went on, broken, endless. Will looked around him vaguely. In the middle of her relatives, Teresa was frozen, staring at the door, as white as if she were facing a ghost.

Will turned—to see Joanna standing in the doorway. She saw Will with Mankau and crossed the room without looking at Teresa; she seemed eerily calm as she looked from the surgeon to Will.

"Sydney . . ."

Will reached for her hands. "She's out of surgery. She's stable."

Joanna looked at Will with no visible reaction, then her knees buckled as she fainted. Will caught her, holding her up.

CHAPTER TWENTY-ONE

Streaks of pink and silver dawn shone through the window in
the hall.

Will and Joanna sat by Sydney's bedside in a curtained cubicle
in surgical recovery. Joanna held her through the IV and oxygen
tubes, seeming to will her to breathe.

Will watched his wife's face, wondering things too strange to
speak aloud. He smoothed his hands on his thighs and felt something in his pants pocket. After a beat, he reached into the pocket
and drew out the black glove.

He sat for a moment with the velvet softness crushed in his fist,
proof that he hadn't dreamed, that he wasn't mad. Then he
opened his fingers and handed it across to Joanna. "You left it in
the garden."

She looked across at him. Her face was very still, but she met
his eyes. "I couldn't just sit . . . waiting . . ."

They looked at each other. Will didn't move, didn't breathe. He
knew that Joanna was on the verge of opening up; she had something she wanted to say. It was inconceivable to Will what that
might be.

Then, between them, Sydney stirred on the bed.

"Mommy?"

Joanna broke from Will's gaze, leaned in to Sydney with frightening intensity. "I'm here, baby."

Sydney looked over at Will and half smiled, drowsy. Will was flooded with relief so profound, he felt light-headed. His eyes filled with tears. He stooped by the bed.

"Hi, bunny."

Sydney lifted her head for his kiss. He touched her face; Joanna held her hand tightly.

"How are you, baby?"

Sydney looked up at Joanna. As Will watched, mother and daughter gazed at each other with that strange connection they sometimes had, a long, odd moment. Sydney spoke with Joanna's eerie composure. "Better."

Will looked at Joanna, then at Sydney. Joanna stroked Sydney's cheek. "Try to sleep, bunny."

Will stepped into the lavender corridor outside. The sense of unreality was stronger than ever. At the sound of footsteps, he looked down the hall and saw Mankau with a group of doctors, speaking in low voices as they hurried along the hall. Connor was not among them.

Mankau slowed as he saw Will. Will stepped up to him. "How is she?"

Mankau didn't quite meet Will's eyes as he answered. "She is stable. We will . . . hope for the best." And as Will watched, a bead of sweat trickled slowly down the surgeon's forehead. Mankau turned quickly and kept moving with the clump of doctors, away from Will.

Will stood still, his pulse spiking with alarm. The surgeon had always been a pillar of calm. What could have him sweating like that?

A voice spoke behind him. "Will?"

He turned. Joanna stepped out of the recovery room. To his enormous surprise, she smiled . . . the first time he'd seen her smile in ages. Her face was dazzling, full of love and light.

"She's asleep. I thought I'd go to the playroom and get some DVDs for later."

Another shock. *Leave Sydney? After she'd only just opened her eyes?* He stammered, completely off balance, "Don't you—want to stay close?"

Instead of answering, she stepped up and kissed him. Such a simple gesture, but from a different world. Will was dizzy with the feel and smell of her . . . he ached to simply sink into the moment. But the alarm bells in his head were now at full volume.

He pulled back and took Joanna's hands, looked at her probingly. She was radiant, almost unrecognizable. He spoke carefully, afraid to break her. "Joanna, all Mankau would say is that she's stabilized."

"I know," she said lightly. *Lightly.* "But she looks better."

She smiled again, a blaze of pure joy, and moved off down the hall. Will stared after her until she disappeared into the elevator.

He looked down through the window beside him at the snow-covered garden, the stone bench, the willow tree . . . the statue of the angel.

He pulled the heavy purple curtains closed behind him as he stepped back into Sydney's cubicle. Sydney lay sleeping on the raised gurney, her breathing even, the low blip of the heart rate monitor strong and steady. Will was fuzzy from lack of sleep, but he studied his daughter intently.

She *did* look better. She seemed less pale, even less thin—her face rounder, her cheeks rosy.

Her eyes fluttered, and she looked up at him, her expression bright and alert.

"Daddy."

Will bent in to her. "How's my girl?"

"Starving," she said definitively.

Will's heart skipped a beat; the imperious tone was so much the old Sydney. But starving? She'd always been violently ill for at least a day after anesthesia.

"It's not good to eat yet, Princess. We don't want you to get sick. How about some ice chips—"

"I want pizza," she said, and added out of nowhere, "I'm going to be home for Easter eggs."

Will looked at her, so startled that he thought he must have heard wrong. "Well, baby, I don't know. . . ."

Sydney gazed back with clear gray eyes. "Mommy said."

By the time Joanna returned with DVDs and board games, Sydney was pushing away the ice chips Will offered her and demanding blueberry pancakes. Joanna promised all the pancakes she could eat as soon as the doctors said so. Joanna seemed to be floating. She was sparkling, even laughing, as the three of them played Sydney's favorite, Chutes and Ladders, on the bed.

Later, when Joanna and Sydney were napping, Sydney curled in Joanna's arms, Will walked into the geometric steel-and-glass wing of doctors' offices.

He went straight to the receptionist. "I'd like to talk to Dr. Connor." Will had tried paging him twice already, but the young resident wasn't answering.

The receptionist smiled up at him. "Dr. Connor's no longer with us."

Will stared at her. "*Connor*—the resident on Sydney's team."

"Dr. Connor's taken a position in another hospital," she said cheerfully.

Will was almost too stunned to speak. "But he was in surgery *last night*—"

"The position he's been waiting for just opened up in New York."

Will wanted to shout, to shake her. He struggled for calm. "He never said anything—he's already gone?"

"That's how rotation works," she assured him. "I'm sure he'll contact you as soon as he's settled in. He's very fond of Sydney. Oh, and Dr. Mankau asked me to tell you they're ready to release her from post-op."

The phone rang beside her, and she reached for it. Will stood, confounded.

CHAPTER TWENTY-TWO

The noise in Will's head was at fever pitch as he stood in the elevator, descending toward the Children's cancer ward.

The elevator stopped midfloor. Will waited, with the familiar sinking sensation . . .

. . . then the doors in both front and back slid open at the same time . . .

. . . the corridors outside both doors elongated into infinity . . .

In the far distance, a woman screamed and screamed . . .

Will woke with a start. He was in bed, in the bedroom of his residential suite.

He sat up, shaking loose from the dream. Connor's sleeping pills were on the bedstand beside him. As his heart returned to normal pace, he realized with a sick jolt that he had no idea what reality was anymore. Was Sydney out of surgery? Had Joanna disappeared?

His dream of the elevator rushed back to him—the woman's screams . . .

He threw off the blankets and bolted from the bed.

———

He ran through the hospital corridors toward Sydney's room, caught in a rising tide of panic. Receptionists, nurses, and visitors turned to watch him as he blazed by. Flashes of memory played in his head: The rotunda. Teresa in the garden. Joanna's black glove.

He strode in the yellow halls past the Easter mural, tore through the doorway of Sydney's room.

It was empty, the bed stripped. Will rocked on his heels at the sight, thinking the unthinkable. He stumbled toward the bed . . . then swung around at movement behind him . . .

Joanna walked in from the bathroom, holding Sydney wrapped in a towel, damp and flushed and glowing. She squirmed impatiently in Joanna's arms, all her former listlessness gone. "Daddy! They have *Princess Mononoke* today!"

Will caught his breath. With a relief beyond measure, he reached out to take Sydney, but Joanna pulled back, crossed to the bed, and stood Sydney up on it. "It's fine, I've got her."

And for just a moment, Will had the incomprehensible sensation that his wife did not want him to touch their daughter.

He had slept through the morning. There had been tests. Instead of Connor, Dr. Mankau came in himself to report to Will and Joanna that Sydney's kidney function was fine; that bladder function was fine; that her vitals were more than fine.

The surgeon spoke to Sydney directly, with grave politeness, as he listened to her heart and lungs, removed the gauze to check over her incision. Will watched him closely. Mankau's casual manner did not quite mask his hyperattentiveness to Sydney.

"So Dr. Connor is gone?" Will's eyes were fixed on Mankau as he asked. He thought he saw the doctor stiffen, almost imperceptibly.

Then the surgeon spoke while continuing his exam, fingers gently palpating Sydney's abdomen.

"Yes, he has been waiting for this transfer for quite some time. We shall miss him."

"We will, too," Will said pointedly. "I would've liked to say good-bye."

Mankau straightened, turned from the bed to face Will. "He was needed immediately. Certainly he will be in touch when he is settled."

He did not hold Will's eyes but looked to Sydney. "Young lady, you are the brightness of my day. Either I am a brilliant surgeon, or you are a brilliant patient."

Sydney giggled, a sound too rich and real for the hospital. Joanna laughed, too, infectiously. Mankau half bowed. "I shall expect no less tomorrow." He turned for the door.

Will glanced toward Joanna and followed the surgeon out into the hall. "Dr. Mankau . . ."

The Indian turned. Will caught up with him. "There's another staff member who seems to have disappeared last night. His name is Salk." His eyes were locked on Mankau's face, his own breath suspended in his chest, but he could see no flicker of recognition, only gentle puzzlement from the surgeon.

"I know no staff person by that name," the doctor said politely. "I am sure administration would be able to help."

Will found Cass in the yellow circle of the nurses' station, writing in a chart. Her broad face warmed as she saw him.

"Good news today, Mr. S. Thank God." Was it Will's imagination, or had she subtly emphasized the last?

He spoke carefully. "We're very grateful."

Cass's smile faltered at his tone. She looked down at her chart. "How is Mrs. Sullivan?"

Will felt a prickle of unease. Cass was always so much more formal with Joanna than she was with Will, but this time there was definitely an undercurrent in her voice.

She lifted her head and met Will's eyes. Will moved closer to the desk. "Cass. Have you seen a man around the ward: tall, elegant, always well-dressed, talking to the patients?"

The big nurse smiled slightly. "You mean other than yourself, Mr. S.?"

Will didn't return the smile. "He said he was a counselor."

Cass hesitated, spoke slowly. "No, I never saw a man like that."

Will studied her intently. "Cass, are you sure? You need to tell me—"

Cass's eyes flashed. "I haven't seen that man you say." Her face closed, and she dropped her gaze. "I hope—Sydney's going to be just fine."

Will walked down the yellow hall past the Easter mural and stopped in the doorway of Sydney's room.

Joanna and Sydney were on the floor with stuffed animals all around them, both bright-eyed, laughing, singing.

Will had not seen Sydney out of her bed since—he couldn't even remember when. Her cheeks were impossibly round and rosy, as if a month had gone by since the surgery instead of just a day. The two of them sang, giddy and giggling.

> *Merrily I prance and sing*
> *Tomorrow will a baby bring.*
> *Merrily I prance and shout*
> *The name the Queen cannot find out—*

Neither of them saw Will as he watched them from the doorway. He was shivering as if he would never be warm again.

———

By dinnertime, Dr. Mankau had cleared Sydney to eat, and she was ravenous. Joanna dispatched Will to the cafeteria for pizza. He walked in the halls in a different kind of daze. The corridors seemed bright and bustling.

In the cafeteria line, Will piled pizza slices on his tray. The bright clattering of silverware and the layered smells of tomato sauce, macaroni and cheese, and warm berry pie were so completely normal, Will was having trouble processing.

In front of him, a dark-haired young man turned away from the cash register, and Will recognized Mark. The slim young man's face lit up when he saw Will. "Hey there. I heard about your little girl."

Will was startled. "You heard—what?"

"You know, close call, better today. I swear, nurses are worse gossips than actors. It's great news," Mark added sincerely.

Will nodded thanks, asked carefully, "How is Gary?"

Mark brightened with hope. "T cells up yesterday." He crossed fingers on both hands, then his wrists for good measure, almost spilling his Coke in the process.

"Good. That's good." Will had a flash of Gary in his hospital bed, with Salk sitting intently by his side. He hesitated, then plunged ahead. "I'm trying to find a man who's been visiting Gary. He said his name was Salk."

Mark looked puzzled. "I don't know who you mean. A man was visiting Gary?"

Will kept his voice level, neutral. "He said he was a counselor."

"Gary's counselor is Dr. Lillian. A woman," Mark added helpfully.

Will felt a mounting paranoia. Mankau, Cass, Mark—had *anyone* seen Salk? He asked the next question without thinking. "The night Gary came into my daughter's room, he said he was looking for someone, a man. . . ."

Mark suddenly looked resigned and tired. "Oh. Well, Gary's mind . . . you know." His voice dropped. "He's been seeing things."

Will echoed numbly, "Seeing things . . ."

"This place," Mark said, and shivered. "This place."

That night as Joanna dressed Sydney for bed, Will stepped out onto the balcony where he had spoken with Salk. He stood at the rail, looking out over the labyrinth of hospitals. His gaze drifted down to the snowy garden, the line of statues.

His eyes found the bench under the willow tree. The bushes moved slightly in the chilly wind. . . .

The glass door opened behind him. Will spun—

Joanna stepped out into the dark with him. Will reached for her hands, looked at her intently in the faint moonlight.

"Joanna. Did you see him?"

Her face was still, unreadable.

"Did you talk to him? Because if not . . . I'm going crazy."

She was silent in the darkness. She turned away to the railing, said softly, "We were both crazy. This place will make you crazy."

She turned back, leaned into Will, and kissed him, lingeringly. Will breathed her in . . . kissed her mouth, her eyes . . . instantly drunk on her, as always, always. He felt her under his hands, taut and yielding . . . always new and molten . . . his other half, his soul . . .

She murmured against his neck, "It's all right. It's all right, now."

Ages later, they moved back into the pale yellow corridor, leaning against each other, bodies fused. Will was floating, detached from his body and all reality. Joanna stopped outside Sydney's door and kissed him again. Before she drew back, she said against his ear, "Sweet dreams."

She looked back at him from the door . . . before she stepped into the room. As if they had a life. As if—everything were possible again.

Will stood in the hall, shaky with wonder . . . and hope.

CHAPTER TWENTY-THREE

I n a few days, Sydney was leaving the room, taking little prom-
enades in the halls, going to the playroom for the *Lilo & Stitch*
DVDs and the afternoon entertainment. Will followed her and
Joanna around in a state of suspended animation, unwilling to say
aloud or even think anything that would upset the delicate web of
optimism and normalcy supporting them all. Instead he lived one
minute at a time, never fully breathing, while Joanna took control
of their daily routine.

This morning, frail children sat around the little band shell of
the playroom, eyes riveted to the stage of a puppet theater, where
Hansel and Gretel puppets walked through a dark, spooky forest,
clutching each other, shivering through their felt bodies.

Will stood with Joanna in the circle of parents, watching Sydney
with the other children, unable to take his eyes off her. She was in
stark contrast with the others, bright-eyed and lively as she whispered with the gaunt little boy beside her. With an inexplicable
sense of urgency, Will moved a little forward to hear, in time to
catch her last words:

"... all gone ... took it out."

The floor suddenly seemed unsteady under Will's feet. Had he really heard that?

The little boy whispered back, "Where can I find the . . ."

As Will strained forward to hear, Joanna suddenly stepped up behind Sydney, whispering, "Shh, baby. Watch the show."

Will stepped back. After a moment, he turned and left the playroom.

He found Dr. Mankau in radiology, a silhouette in the ghostly light of the X-ray viewing lab. The surgeon turned from a light board as Will pushed through the doors.

Will stood facing him. "We need to talk."

Mankau hesitated for the slightest second, framed against the wall of skeletal images, then he nodded. "I agree."

They gathered in Mankau's office again: Sydney's team, minus Connor, clustered uncomfortably around Mankau's desk; Joanna sitting beside Will on the black leather couch in perfect stillness. The animal gods watched from the tapestry on the wall.

As the last doctor took his seat, Mankau stood from behind his desk. "Well. It seems the experimental treatment we started last week has had a startling effect. I noticed it the other night in surgery, but we needed the confirmation of MRIs and blood tests . . . and the latest X-rays are unequivocal."

He turned on the light board under an X-ray of Sydney's tumor. Will flinched as always at the familiar, grotesque shape. Mankau slipped a second X-ray onto the board. Joanna didn't move. Will stared, at first unable to process what he was seeing. The image was taken from the same angle, but the tumor was half its previous size.

"Quite an impressive reduction, as you can see." The surgeon's usually tranquil voice seemed too hearty.

Will looked to Joanna, then to the doctors. "I don't understand."

"Sydney is in remission."

Will stared at Mankau, stunned by the words they'd never thought they would hear. The surgeon was speaking of cell counts and tumor density, but all Will could hear was another voice: *I have seen astonishing things happen. . . . If one miracle has ever happened in the world—why not for you?*

Joanna spoke suddenly on the couch beside him. "Does that mean we can take her home?"

Will looked to her, even more startled. Mankau glanced at the other doctors. There was a beat of silence, and then he replied, "In fact, the cancer is responding so well to the new drugs, and Sydney's vital signs are so good, that we feel there is no reason not to continue treatment on an outpatient basis."

Will was hit with a wave of disbelief so overwhelming that for a moment he couldn't speak. He stared around at the doctors. "Home? You mean . . . but how—can that—?"

Joanna spoke beside him. "Will . . ."

He turned to her. She was shaking, and then her tears were falling, softly, silently, her face lit from within.

"Will . . ." She reached out to him, moving into his arms, and he was pulling her against him . . .

. . . and brushing at her tears, trying to contain his own . . . the two of them in a fog of happiness and wonder, oblivious to the doctors.

He tightened his arms around her, and she shivered with sobs against him, clinging to him . . .

Completely there . . . completely his.

CHAPTER TWENTY-FOUR

It didn't happen that very day. There was a whole file cabinet's worth of paperwork to sign, outpatient treatments to set up, another battery of tests.

In the end, it was Joanna's happiness that carried Will through, that prevented any ominous questions from surfacing. For what, after all, could he say? That Sydney was *not* in remission? That he had seen Mankau sweat the morning after surgery? Will hadn't believed any kind of recovery possible, so it only made sense that the doctors were caught off guard as well. The word *miracle* floated at the edges of his mind; he didn't let it any closer than that.

Packing was a dreamlike experience, the lightness of elation. Dressing Sydney in street clothes and shoes for the first time in months—the realization that they were taking her home, alive. Will hadn't fully let himself understand that he had expected to see his daughter's corpse lying in the hospital room, that he had been preparing to put her in the icy ground.

Instead, they were dealing with how to pack three dozen toy rabbits. Sydney refused to consider leaving even one, and Will griped and sighed, martyred, about hiring a moving company, about building a new rabbit wing of the house. It was fake complaining, a complete sham, and nothing had ever felt so good.

At one point during the rabbit migration, Will sensed someone hovering in the doorway, watching from behind. He turned quickly, not admitting to himself whom he feared he would see.

But it was Cass watching them, with something strangely ambiguous in her face. Will's heart dropped. Then Sydney saw Cass and ran to her, and Cass opened her arms and scooped her up in a huge hug, and Will's unease melted.

"You gonna come back and visit me?" Cass demanded sternly.

"Uh-huh," Sydney replied, muffled against the nurse's ample bosom.

"Well, you better, or I'll come find you my ownself."

But when the big nurse said good-bye, she didn't quite meet Joanna's eyes.

The hospital playroom was more crowded than usual when they stopped by on the way out, bags of rabbits in hand. Will looked over the room, the group games, the coloring stations, parents hunched at the tiny tables and chairs with their children.

Will and Joanna stayed back by the doorway and let Sydney go ahead on her own.

The children stopped playing as they saw her. A hush fell over the room. Sydney walked toward them.

Will watched, riveted, as the children gathered slowly around her. She seemed from a different dimension in her street clothes and coat, the bloom of health already on her. Parents scattered around the room looked at Will and Joanna with undisguised envy—some happy for them, some wistful, some with desperate, palpable hatred.

Will shifted uneasily under the bombardment of raw emotion. Beside him, Joanna was calm and still.

Sydney was speaking solemnly to the circle of children. Will focused in to listen.

"I won't be back," she said with such authority that Will felt a chill.

Around her, the other children nodded solemnly, sagely. Before Will could react, Joanna was right next to him. She held out a hand to Sydney. "Baby . . ."

Sydney walked to Joanna, put her hand in hers. They turned and walked away from the circle.

Will looked back at the circle of children. They watched, silent and still, as Sydney left them.

Dr. Mankau was waiting for them in the hospital lobby. "There's my girl," he said gravely to Sydney.

Joanna nudged her forward. "What do you say, baby?"

Sydney walked up to the doctor, and he crouched to her level. "Thank you, Dr. Mankau."

He took her hand in his. "The pleasure was entirely mine."

He stood, and Joanna stepped up. And surely it was Will's imagination that the surgeon also did not quite meet Joanna's dark blue gaze. "We owe you everything," she said simply. Mankau bowed over her hand.

Will hovered behind with the doctor as Joanna steered Sydney toward the desk nurse for more good-byes.

"Are you sure . . ." he started.

The Indian looked him square in the eyes. "I have been practicing medicine for twenty-two years now. Long enough to know that sometimes, we get handed a miracle. This time it was for you. My professional advice is to thank God and grab it with both hands."

Will nodded, gripped Mankau's hand.

Then he joined Joanna and Sydney, and they walked out into the living world.

The pneumatic doors whispered shut behind them.

CHAPTER TWENTY-FIVE

The day was a blur of happiness . . . the hospital complex and the crush of eager press receding in the distance as Will drove the BMW through the Boston streets. Sydney sat up wide-eyed and excited in her car seat, pointing out new and familiar sights: the park outside magnificent Trinity Church, the rows of gold-wrapped bunnies in the windows of the Lindt chocolate store, the huge bronze teddy bear in front of FAO Schwarz—and Sydney's favorite place "in the *whole world*"—the Public Garden. She insisted on stopping, and how could they deny her?

Overnight, winter had turned to spring; the light was so clear and bright, it hurt Will's eyes. The snow was gone. In the park, trees and flowers were bursting with blossom; robins and seagulls and squirrels shared the paths with skaters, dogs, lovers, parents with strollers.

On the clear lake, barges with swan figureheads drifted under flowering trees, and in no time Sydney stood at the head of one like the Lady of Shalott, regally scattering bread crumbs to ducks.

Will and Joanna watched her from their seats on the barge—their miracle. Joanna leaned her head against Will's shoulder; he kissed her hair. Then, looking out past the bank of the lake, Will saw the little bandstand where he'd given his speech. His face grew intent, distant . . . in his mind, he heard the faint sound of applause.

He felt Joanna watching him, all-knowing, always.

The sun went down in fiery glory outside the Tudor house in the woods with its long, curving drive, the garden thawing, waiting for Joanna's magic touch.

Unbeknownst to Will, in the middle of everything, Joanna had called their housekeeper to air out and spruce up the house; they arrived home to lit rooms, a crackling fire in the hearth, Joanna's tortoiseshell cat, Salome, curling around their feet in greeting, and the fragrance of fresh flowers everywhere, bouquets and baskets and extravagant arrangements of well-wishes sent from all corners.

The light in Sydney's bedroom was warm and glowing, the room was filled again, alive, with Sydney's things unpacked . . . stuffed rabbits and dolls and books crowding the shelves.

Squeezed into Sydney's little bed, Joanna and Will and Sydney read aloud from *The Blue Fairy Book*. Will and Joanna sipped red wine from long-stemmed glasses, held the book between them for Sydney. The cat nestled purring at their feet.

They looked at each other over the top of Sydney's head as she read.

". . . and Prince Darling brought the Dear Little Princess out of the dark forest to his palace . . ." Sydney turned the page and looked to Joanna for her turn.

". . . and there was great celebration throughout the land. And if they have not stopped celebrating, they are celebrating still." Joanna paused dramatically. "And they all lived . . ."

Sydney and Will shouted with her, those magic, incredible words, "Happily ever after!"

Joanna closed the book. Sydney settled down into her pillow, sleepy, content.

"My room's bigger."

Joanna slid off the bed to kneel and hug her. "Sweet dreams, baby."

And Will bent to kiss her. "Be good."

"*You* be good," she retorted, an old game of theirs. Tears threatened behind Will's eyes.

At the door, he and Joanna stopped to turn out the light. They leaned against each other, arms tight around each other's waists, looking at Sydney in the glow of the night-light. The happiness kept coming in waves. Wonder. Euphoria. Will was dizzy with it.

And as they turned away from Sydney's doorway, they were in each other's arms. Mouths seeking, clinging, hands all over each other.

Joanna leaned back against the wall . . . Will pressed into her, kissing her neck . . . her hands slid under his shirt, baring skin. She breathed into his ear:

"Be good."

They made love in their own bed, one flesh, with nothing between them, gasping, transported, on fire.

The waning moon crept into the dark sky and watched through the windowpanes behind them.

CHAPTER TWENTY-SIX

Will awoke alone in bed. For just the slightest moment his heart fluttered, a memory of fear. Questions that needed to be anwered, questions of life and death . . .

Then he heard the murmur of feminine voices downstairs and birds chirping in the trees outside.

He opened his eyes to sunlight streaming through the windows. He was in their own room, in their own bed, under silken sheets, with Joanna's smell in his hair, the memory of her body on his skin.

Everything felt so *real*.

And he thought, *Yes. I choose this.*

He banished the hospital from memory, stretched lazily . . . and smiled.

Downstairs in the kitchen, Joanna stood barefoot at the stove, making pancakes, looking tousled and sexy in a short, faded paisley robe Will refused to let her get rid of. Sydney knelt on a chair at the kitchen table, dipping eggs into colored dye, a huge beribboned basket of Easter chocolate beside her, glitter all over the table and floor around her. Salome ate from a bowl at Joanna's feet.

Will stood in the doorway, breathing in the smell of coffee and blueberries, marveling at how extraordinary normal actually was.

Joanna looked up from pouring blue-spotted batter; the look she shot him was purely pornographic as she said innocently:

"Daddy's a sleepyhead."

Sydney waved him over. "I'm making eggs," she announced.

Will gave Joanna a sizzling look back before he leaned over Sydney, kissed the side of her mouth. "Mmm. Chocolate."

Sydney giggled, brushed him away. She reached for a waxy crayon. Joanna turned from the stove, watching them. Will looked at her over the top of Sydney's head.

"Happy Easter," Joanna said softly. Will was startled—the date hadn't even occurred to him. Resurrection, rebirth, new life . . . the symbolism gave him a momentary chill of wonder.

Joanna moved back to turn the pancakes. Will pressed up behind her, slid his arms around her. She closed her eyes, leaning back against him with a sigh. He discreetly pushed down her robe, nuzzled the delicious softness of her neck. As his lips touched her shoulder, she shivered. Will felt the flinch and opened his eyes—to see a livid purple bruise in the hollow of her neck.

A little shocked, he mouthed, "Sorry," and kissed the bruise gently.

She touched the bruise unconsciously. "I enjoyed it." Her face was blank as she turned back to the stove.

Sydney held up an egg. "Daddy, you do one," she commanded.

Joanna glanced over her shoulder. "Uh-uh. Later. You two need to get dressed."

Will looked at her from the table. She pushed back her hair with a hand, holding the spatula in the other.

"Church. And then Senator Flynn's Easter picnic after."

Will was too surprised to speak. There was only one reason

Joanna went anywhere near a church, and that was for him, for the politics of it. And Flynn's Easter picnic was pure politics, a fund-raising extravaganza. Will had never missed it, and of course had planned to go this year, but that had been . . . before. He felt a twinge of unease. Surely Joanna didn't expect him to go back to the campaign?

He glanced toward Sydney, who was shaking another cloud of glitter onto an egg, and lowered his voice. "Joanna. Let's take this one day at a time—"

She turned on him with startling intensity. "We're *fine*." She stood tensed, the spatula trembling in her hand.

Before he could say anything more, the doorbell rang in the front hall. "That'll be Jerry," Joanna said casually. Will looked at her, not understanding. She stepped to a cabinet, took down two sunflower coffee mugs. "I thought you should talk before you see Flynn today. You have a lot of catching up to do."

Off balance, Will looked again toward Sydney, lowered his voice. "Joanna, we don't know—"

"*I* know," she said violently. Will looked at her, unnerved. The doorbell rang again, and she smiled then, but it was, he thought, a little forced.

"Go talk. I'll bring in coffee. We're fine."

In Will's study off the living room, Jerry upended a briefcase, dumped it onto the mahogany desk. Rubber-banded stacks of cards and letters spilled out, covering the desk. Will looked over the piles: e-mails, greeting cards, handwritten letters. Jerry walked the carpet, practically bouncing with enthusiasm. "And those are nothing. Rooms full of it. And money to match."

Will stepped to the wide window, looked out. In the garden, Joanna and Sydney were picking flowers to go with their lacy Easter dresses. The sun was shining; everything was green. . . .

Jerry spoke behind him. "You had a miracle. Everyone wants to be part of it."

Will turned from the window. "I dropped out."

"Never official," Jerry said quickly. He picked up a stack of letters, riffled through it. "Just read what some of these people are saying, Will."

Will glanced down. A phrase popped out from a card: "God has chosen you. . . ."

He looked up at Jerry, his throat tight. "I just got my daughter back. I'm not going to start campaigning—"

"You don't need to *do* anything." Jerry found the remote on an end table and zapped the TV on to a cable station. He didn't even have to flip through channels to make his point. On the screen, Will saw images of himself and Joanna taking Sydney from the hospital, the press crowding in around them.

"The whole country's behind you," Jerry said, watching Will's face. "If we held an election this afternoon, it would be a landslide."

Will stared at his family on the screen, at Sydney's face, full of light. Jerry spoke behind him. "It's yours, Will. All you have to do is take it."

CHAPTER TWENTY-SEVEN

It was a circus. Something just short of the Second Coming.

Holy Eucharist services were at Trinity, the Back Bay's lavish Romanesque church built by Henry Hobson Richardson in the 1870s, a castle of rough-faced stone, heavy rounded arches, and a massive central tower.

Valets parked luxury cars in side streets along Copley Square while Boston's first families milled on the wide church steps and the plaza outside, dressed to the nines in Easter finery. The trees in the plaza were an explosion of blossoms, white, pink, purple: magnolias, pears, cherries.

Will had experienced the circus ever since he could remember, but this was beyond anything he'd ever seen before. Photographers and journalists swarmed in full-out frenzy, and the crowds—where had they come from? Families holding flowers, throwing blooms like attendants at a royal coronation. Thank God Jerry had been thinking ahead; he'd had a whole firm of bodyguards waiting outside the house to escort them to the church. Now large men in suits muscled through the waves of gawkers, controlling the flow of people around the family.

Sydney was in heaven. Never a shrinking violet where cameras were concerned, she seemed to understand perfectly that *she* was the main attraction. Reporters called out all around them: "Sydney, look this way, honey . . ." And she smiled beatifically into the lights and nodded and waved with regal composure.

Joanna hovered, a vision in watercolored silk and lace, never an inch away from Sydney but allowing her to take the spotlight. Will wondered briefly that she would subject Sydney to this kind of display so soon after they had reclaimed her, but of course she couldn't have known any more than Will had. Will himself was staggered by the outpouring.

Then Senator Flynn swept across the top steps to greet them. Cameras flashed, with pops of bright light . . . and the crowd literally parted on both sides to let the Sullivans through—a classic photograph. Will knew it would be splashed all over the papers the next morning. It might even make *Time*.

Inside the cathedral, the congregation rose as one to sing under the enormous dome of the church. Intricate stained glass broke the light into jewels.

Will reached for a hymnal, though he knew the words ("Christ the Lord Has Risen Today") by heart. He didn't quite have his father's classic Irish tenor, but he'd always been able to make ladies of a broad spectrum of ages sigh. He lifted his voice to join the others echoing off the cool, curved walls. Trinity was his mother's church, the show church—magnificent, yet more fleshy than ethereal, with its interior so deeply, surprisingly red. The wooden box they stood in had been his mother's family's for generations; it was all familiar, from the musty smell of the dark cherrywood pews to the embroidered kneelers at their feet bearing their own names. Sydney stood on the bench between him and Joanna, her hand on Will's shoulder, Joanna's arm around her.

Will looked up into the ribbed dome of the chancel. His gaze was caught by the brilliant stained-glass window directly above them: La Farge's classic Resurrection scene: *Mary Magdalene Before the Empty Tomb*. It hit him that he was in church, with his daughter, who would have been dead, standing alive beside him.

A miracle. A miracle.

But as he stared up at the veiled woman in the window, the gaping black mouth of the tomb, he felt not elation, but a thrill of horror, as if a tomb had opened beneath his own feet.

Will pulled his eyes from the dome, turned in the redness of the church . . . and saw that beside him, Joanna and Sydney were just standing quietly.

Not singing.

And he knew in that moment he *didn't* believe—not in the song he was singing, not in miracles, not any of it.

Around them, the music drifted heavenward.

CHAPTER TWENTY-EIGHT

Children raced over the immaculate lawn and blossoming gardens of Senator Flynn's sweeping Brookline mansion. Caterers circulated between luncheon tables under a blinding white marquee; elegant, connected guests sipped champagne as they watched the children wandering with baskets, searching the flowers and bushes for hidden Easter eggs and chocolate bunnies.

Will found his bleak moment in the church melting away in the sun, and he let it go gratefully. He and Joanna stood on the new grass with arms around each other, watching the great hunt . . . laughing as Sydney dashed up to show them her latest find.

Flynn's Easter picnic was a Boston institution, as crucial to the political calendar as the St. Patrick's Day feast that unofficially kicked off the Massachusetts campaign season. Will hadn't been more than eight when he'd begun to realize that though the children's egg hunt took center stage, the true purpose of the picnic was the adult maneuvering.

Now he stood with chilled Veuve Clicquot in hand, an arm tight around Joanna's silk-draped waist, and was almost disappointed to find his political mind returning . . . observing clinically that though no press was in evidence, photos of Sydney racing around the lawn with her loving parents prominent in the background would miraculously appear in all the papers. *A good ten polling points right there.* He thought again of *Time*. It could even be the cover.

A discreet distance behind them, a clump of politicos hovered with Jerry and Senator Flynn. Just a few old friends, but power emanated from them in waves. The kingmakers watched Will and Joanna and Sydney greedily. Will didn't have to hear to know what they were saying:

Couldn't ask for that kind of publicity. A fucking medical miracle.

Snatched from the jaws of death. Truly, almost mythic. Will could see it from apart as if he were not a participant. If he chose to run, he was absolutely invincible now. Benullo had always hated Will, his idealism and charisma, his golden pedigree. Now he must be quaking down to his Ferragamo shoes.

The change in fortune was dizzying. Surreal. As if the long nightmarish winter had never happened. . . .

No. As if it all had happened—and the result was an elevation to undreamed-of heights.

Will saw Jerry raise a hand to flag him over. Will's face shadowed. But he kissed Joanna and moved across the lawn toward the clump of political insiders.

He joined the men, shook hands all around. A businessman he recognized as a campaign contributor pumped his hand energetically. "Believe me, you had the prayers and well-wishes of a significant number of people in this state."

Emphasis on the *significant,* Will noted, not aloud.

"We're up twenty points on Benullo in tomorrow's *Globe* poll," Jerry said.

Oh, they can taste it, Will thought. It wasn't like Jerry to be so unsubtle. Will said nothing. The other men shifted on their feet.

The sun seemed suddenly too warm on their suits; Will could see sweat at several elegantly graying temples. He waited the men out. Finally another spoke.

"In light of the . . . new circumstances, have you been giving any thought to running?"

Will allowed himself a grin. "The only running I'll be doing for a while is after Sydney."

The politicos laughed smoothly. One of them had the gall to clap him on the back. "Couldn't ask for better practice."

Will smiled, perfectly pleasant, but his eyes kept drifting to Joanna . . . lively and radiant as she knelt with Sydney to exclaim over her chocolate treasures. Sydney's head was tipped back as she laughed, all little-girl sunshine.

Another contributor, a dairy man, spoke. "Got your priorities in order. That's exactly why this country needs you."

So it's the country *that needs me now,* Will thought, and was ashamed that the thought was thrilling.

Flynn raised a hand, acutely aware of Will's discomfort. "Enough shoptalk. It's a holiday."

"It certainly is," Will agreed. "Excuse me." He walked away from the politicos, across the perfect green grass toward Joanna and Sydney.

All at once he felt eyes on his back, as real as cold fingers touching his neck. He turned slowly.

The cluster of politicos was still watching him . . . but behind them, against the brilliance of the sun, he caught a glimpse of a tall figure with chiseled cheekbones, dark hair . . .

Will tensed. It couldn't be . . .

But he strode forward toward the crowd anyway, his face tight, his eyes searching behind the clump of men.

Then out of the corner of his eye, Will saw a door close in the house behind. He veered on the lawn and walked quickly toward the side of the house, weaving on fieldstone paths through the immaculate rose garden, in pursuit. The floral scent around him recalled the hospital; Will flinched and tried not to breathe.

He came up on the trellised patio and pushed in through the French doors. He found himself in Flynn's study, a larger, more opulent version of Will's office.

The room was empty.

Will breathed out, untensing slowly. *Chasing shadows again.*

His gaze drifted over the office: a stately leather chair behind a satiny walnut executive desk, built-in bookshelves filled with gilt-edged volumes, French doors leading out to the rose garden that not-so-subtly recalled another Rose Garden. The room reeked of power; it was in the air, like the trace scent of aftershave and cigars.

Will stepped over to a wall of framed photos, the history of a political career. Instinctively, he zeroed in on a photo, his own father's inauguration: a younger, more idealistic Hal with his hand raised; Will's mother, Susan, a fashion plate as always, looking on proudly; Will as a small boy, gazing up at his father as he took the oath of office.

Will suddenly missed his father more than he could say. Joanna was his best friend, his soul; he rarely talked intimately to anyone else. But they hadn't talked, really talked, in months. Despite their explosive connection of the night before, there was a strange distance between them—undercurrents of things that could not be said. The loneliness was almost more than he could bear.

A male voice spoke from the doorway, and Will's hair stood up on the back of his neck, so sure was he of who it was.

"What's wrong, son?"

Will turned. Senator Flynn stood in the doorway, watching him.

For a moment Will recalled another study, his father's, and himself as the new district attorney, having found the proof of what he'd long suspected: corruption, graft, payola from contractors who had built his father's political dynasty. His father slumped at his desk before him, defeated and proudly defensive as he looked up at Will with red-rimmed eyes. *For what it's worth, no one will ever be in a better position than you are. No one will ever start cleaner. You'll never have to make the—compromises—I did in the beginning, because your old man did it for you.*

And Will turning in disgust, walking out. A week later, Hal was dead of a heart attack; the fight had been their last exchange.

Flynn was looking at Will, waiting, quizzical. Will banished the ghost of his father, groped for words. "Everything happened so fast. Sometimes it doesn't feel real. . . ." He trailed off, unable to explain. He moved restlessly to the French doors to look out over the garden. His eyes found Joanna and Sydney on the lawn. "I think I'll wake up back in the hospital."

When he spoke again, there was steel in his voice. "I am not going to exploit my daughter's recovery for politics."

Flynn said swiftly, "No one would ever ask you to do that. My God, Will. I'm a grandfather." To emphasize the point, the senator approached the French doors, looked across the lawn to his daughters—one with her husband, rosily pregnant for the third time, the other staid and maternal, sipping a mimosa, free of her teenagers for the day.

Flynn turned back to Will, his face softer. "But . . . what if all that's happened was meant to be?"

Clouds drifted over the sun outside, and the room darkened slightly. Will watched Flynn warily.

"You've marched into the gates of hell, and survived—with your family intact and stronger. That's the kind of fire that hones great men—and great politicians."

Will felt a chill, an almost trancelike sense of significance. He was aware of the manipulation, but all politicians were magicians. They had to be.

"God works in mysterious ways, Will. Sometimes we don't choose. We're chosen. Are you going to turn your back on what is looking very much like destiny?"

Destiny. Of course. Will felt short of breath. He rubbed his face. "You are good," he finally said.

Flynn laughed, clasped Will's shoulder warmly. "Don't be cynical, bucko." He met Will's eyes with a steady gaze. "And don't ever think I'm not in awe about this. You'd have to be made of stone not to be." He turned and looked across the lawn at Sydney and Joanna with real affection.

"Take some time. Be with them. Just show up at a few fundraisers, give a few speeches. And whenever you want to talk about plans, we're holding your place. Governor's just the stepping-stone, Will. You can go all the way, if you want it."

CHAPTER TWENTY-NINE

Flynn's last words to him rang in Will's ears as he stood in the cool night on his front porch, looking out over the dewy lawn . . . up at the waning moon above the woods. The pronouncement was as good as an anointment, coming from Flynn.

You can go all the way, if you want it.

But Will didn't need Flynn to tell him. He'd lived and breathed politics since he was born. He knew the power of this particular crossroads.

He felt a subterranean pull . . . seductive, voluptuous, disturbing in its intensity. He couldn't deny a kind of predestination in everything that had happened. But that was the Devil's voice— pure temptation. What politician *didn't* think he was called, marked, destined in some way?

Destiny.

What if Flynn was right? What if all this was meant to be?

And if it was a gift—how could he refuse it?

He heard a step behind him and turned to see Joanna standing silhouetted in the light of the doorway. She came to him, slid her arms around him from the back, burying her head on his shoulder, smelling of wine and lilacs. Without preamble, she said into his neck, "It's what you've always wanted."

Will laughed. There was nothing he had ever been able to hide from her. He turned in her arms to face her, gathering her close against him, searching her face.

"But it's nothing if I don't have you."

She pulled back a little, puzzled. "But you do have me. You always will."

She snaked her arms around him again, this time with her hands under his shirt, her fingers against his bare skin.

"How can you not run? It's a gift." Her hands moved lower. And Will melted, helpless at her touch.

When he walked into the bedroom, a fire jumped and danced in the fireplace. Joanna stood by the bed in a silky robe, pouring champagne from a bottle on the bedside table.

Her back was to him, and when he moved up behind her and slid his arms around her waist, she flinched but recovered quickly. "Why, Governor," she said in a throaty voice, and turned to face him, offering a glass. But Will was disturbed. She hadn't flinched from his touch since they first knew each other and she was still learning to trust.

Before he could ask her what was wrong, she stepped into his arms, drawing down his head and kissing him deeply, opening her mouth under his, biting his lips. Will tasted champagne and honey. He set his glass on the table and pulled her closer.

Joanna moved back and reached for the light, turning if off . . . just a beat too quickly.

Will frowned.

She stepped back from him. Standing in just the firelight, she slid off her robe, let it drop . . . to reveal an intricately cut-out lace negligee he'd never seen before, breathtakingly sexy . . . and concealing.

She turned slowly to let him see her, her dark eyes shining as she saw the fire in his. She slipped into his arms, and he kissed her hungrily, his fingers exploring the tempting cutouts of her gown.

His hands slid down her silk-covered thighs to lift the hem . . . but Joanna put her hands on his, stopping him. She pushed him slowly back on the bed. Moving like a cat, she crawled over him, stripping off his clothes in a slow tease, keeping hers on. She worked her hips into his, shifting deliciously on top of him. Will dropped his head back, moaning in pleasure, hardening under her.

She stretched her arms above his head, grazing lacy breasts across his naked chest . . . and lifted the glass of champagne from the table. She took a slow sip, then bent to kiss him deeply, gushing the wine into his open mouth.

Alcohol and desire burned through him. All thought gone, he drank the champagne . . . drank her. She was feverish herself, urgent and demanding.

They rolled on the bed—over . . . and under . . . flesh fusing with a primal shock . . . and she was straddling him in the firelight, riding him, her legs were against his thighs . . . he was gasping . . .

She clung to him, driving him harder into her . . . and Will looked up into her eyes. They were strange, fiery, compulsive.

Will was caught, shocked, suspended in the moment . . . then all was erased in the shuddering force of his climax.

CHAPTER THIRTY

So it was decided. He was back on the trail, spending his days on the stages of Rotary Clubs and church assembly halls; taking meetings in the backseats of town cars; writing speeches at the breakfast table.

Joanna wanted it. She had always been passionately supportive of Will's political aspirations, but now it was as if she needed to have everything completely normal again. She had an almost superstitious fervor to go back to exactly where they had been before she found the tumor—to erase those dark months completely from the record of time.

And there was no rational reason not to. Every visit to the hospital, every set of X-rays and MRIs, showed another reduction in the tumor. Sydney was in rapid remission, the cancer cells dying away. And with the cancer died the fear, the anxiety, the doubt.

He had it back—daughter, wife, life, all of it.

He looked out from the platform over a sea of faces as brilliant camera flashes rippled through the fund-raiser crowd laced with photographers and press. Joanna sat behind him, in tangerine linen, Sydney on her lap . . .

The old rush electrified him as he spoke, his own voice rever-
berating around him. "I see Massachusetts—I have always seen
Massachusetts—as the soul of this country. Now, a cynical person
might say I'm prejudiced."

Appreciative laughter from the hall washed over him. Will
grinned in acknowledgment. "But I think—not unjustifiably
so—that the best of the history of this great nation originated in
this commonwealth. Our great leaders and fighters. John Adams.
Patrick Henry. Samuel Adams. Our laws. Our social ideals. Our
Constitution and Bill of Rights, which became the basis for the
great Constitution of these United States."

The applause was deafening, of course. He waited, finally had
to raise his hands to stem the tide.

"We are the oldest yet most progressive; the most historically
rich yet forward-thinking state in the country. And my dream—
my dream—is to see Massachusetts take the lead in this country
again—to remember our ideals, to inspire again by our own example. We need a leader who is up to this task."

He paused, looked hard out over the hall. "I don't believe Governor Benullo is that man." Below him there was some low, surprised murmuring at the direct attack.

"These are challenging times—for our state and for our country. We've been awakened to dark forces, new threats at home and
abroad. Governor Benullo's response is to wave flags. The patriot-
ism flag. The 'family values' flag."

Will spread his hands. "Who can argue with family values?"
He looked back at Joanna and Sydney. Hearty chuckles rippled
through the crowd.

Backstage, Jerry stood with some politicos, watching Will. He
spoke softly. "Try 'Family's Love Triumphs over Cancer.'" The
politicos around him nodded, watching Will like sharks.

Onstage, Will continued, "Well, I can—when that outdated
buzz phrase is a smoke screen for a reactionary political agenda

that has nothing to do with family, or values, or virtue. We can't afford to hide behind the old familiar clichés that make us feel good about ourselves—by making everyone other than us the enemy. We must call on our own resources and not give in to the shadow terrors of our own minds. Our worst nightmares come true when we give in to our fears—and worse, project them onto others."

Will paused to look over the crowd, feeling the silence: listening, thoughtful, intent.

He looked back at them gravely. "These are challenging times. But I hope and believe that my years as district attorney, and as a son of Massachusetts, have honed me for the fight."

He paused, then finished quietly, "I promise you . . . I'll do my best."

And that was the week, and then the month: an escalating series of appearances throughout the state. A women's club, where Will walked off the small stage and circulated among round tables of tastefully dressed women who ignored their crème brûlée and watched Will's every move with appreciative eyes:

"Governor Benullo is not thinking of children when he proposes to eliminate day care. He is not thinking of women when he slashes funding for family-planning clinics, or of the sanctity of marriage when he jumps on the antigay bandwagon. He is bowing to the pressure of an ultraconservative lobby that owns him as surely as I own my car. . . ."

. . . and a motor factory, where Will spoke in his shirtsleeves on the factory floor, surrounded by union workers:

"I have hope—that we, the people of Massachusetts, can do better. That we can give *public* education the funding it deserves . . . guard our pensions against the ravages of corporate greed. . . ."

. . . and a dock with Boston Harbor vast and blue behind him, families gathered on the pier below:

"I have hope—that we can protect ourselves and our country without sacrificing the freedoms this great nation was built upon—the freedoms that originated right here, in our own commonwealth.

"I have hope—that Massachusetts can again become that shining city on a hill; a beacon of light and law for this country and other countries around the world.

"I have hope—that we can come together as one people, as one family—and move beyond fear, darkness, and uncertainty to a new and brighter future that is waiting there for all of us."

He looked out into the crowd and into the eyes of the nation.

"I have hope. Do you?"

At the state Democratic convention in June, Will won an unprecedented 87 percent of the delegates on the first vote, knocking both his remaining Democratic opponents off the primary ballot. The *Globe*'s front-page headline the next day read: SULLIVAN PITCHES SHUT-OUT.

The one independent and the Green Party candidate were running message campaigns, with no hope of election. It was now a two-man race, Will against Benullo in the general election in November. Reeling, the Republicans kicked their money machine into high gear.

Will used the campaign to start to lead, touring farther from home now, in counties from Pittsfield to New Bedford, forging the alliances and relationships he would need to run the state. And it felt right—God, did it feel right. It felt *real*. He was back to what he was born to do. It was time, for him and for Massachusetts.

He was unstoppable. Anything was possible, now.

CHAPTER THIRTY-ONE

*L*ight breaks on a bleak morning, gray and chill. Will sleeps peace-fully in bed.

    Downstairs in the entry hall, the grandfather clock ticks. The tor-toiseshell cat naps curled on a straight-backed chair. The cat suddenly raises its head, looks toward the front door . . .

    The door opens almost without sound, and Joanna slips in, closes it behind her. She is dressed in a black linen sheath and sheer stockings, her makeup smudged under her eyes.

    She stands in the dark hall, smooths her hair with a shaking hand . . . then catches a glimpse of herself in the mirror—and quickly turns away.

    She slips off her shoes and moves noiselessly up the stairs.

Will woke sharply. He looked immediately over at Joanna's side of the bed.

    She was there, sleeping soundly. The day outside was beautiful—bright, clear light . . . no sign of the ominous fog of Will's dream. His tension slowly dissolved. He reached for Joanna, touched her shoulder—

She bolted up in bed with a gasp.

Will pulled back, startled, disturbed. "Hey ..." Joanna stared at Will, the wildness fading slowly from her gaze.

He looked at her intently. Her face was pale; there were deep circles under her eyes. He hadn't seen her so haggard since—the hospital.

The hospital.

Joanna pulled the down comforter around her, an unconsciously protective gesture, and smiled shakily.

"I was dreaming."

Will was familiar with her dreams. The rage for her father boiled up again, but he knew from long experience that he had to tread carefully.

"I had a bad one, too," he said lightly, nothing too intense or probing. "I'll tell you mine if you tell me yours."

Joanna's face was shadowed, but she looked into his eyes. Will held his breath, afraid to move, knowing she was on the verge; the door was opening.

Then Sydney ran in from the hall in bunny pajamas, hurled herself at the bed. "Mommy, Mommy, *up*."

Joanna's face brightened. "Morning, baby. How's my girl?" She scooped Sydney up, biting her neck as Sydney squealed.

Will continued to look at Joanna, but the moment was gone. She tightened her arms around Sydney, shook her head, managed a smile and a shrug. "Don't remember."

She shifted her gaze away from his, kissed the top of Sydney's head.

Will kept his eyes on her but raised his voice for Sydney's benefit.

"Who's up for a trip to the park?"

Sydney bounced on the bed ecstatically. "Me me me ... can we go on the swan boats, Mommy?"

Will caught her on the last bounce, sending her into another fit of giggles. "Just you and me, sweetie pie. We're going to let Mommy rest today."

He reached his free hand to touch Joanna's face. And instead of protesting that she was fine, she looked at him gratefully.

Scalding water pounded on Will's back in the shower. He closed his eyes, bowed his head into the stream. His stomach was knotted with apprehension. He'd been gone how many days that week, that month? How many nights had he left Joanna alone?

He was still brooding as he stepped out of the steamy shower and used a towel to wipe the mirror clean so he could shave.

He opened the hamper to drop the towel in—and stopped, staring down into the basket. He pulled at the corner of a black garment, extricating it from a layer of discarded towels.

It was the black linen dress Joanna had been wearing in his dream. He brought it to his face and smelled it—caught the fragrance of her perfume, and something musty underneath, fetid and sour.

Heart in his throat, he bent, digging deeper through the clothes . . .

. . . and stood, drawing out a silk stocking . . . black and sheer . . . and ripped.

He stood in the dim garage, with its faint smell of gasoline and built-in shelves of neatly arranged tools and supplies that he never had the time to use, looking at Joanna's black Cougar.

Sleek, powerful, gleaming—the car was a throwback to her childless days, when Will would sit white-knuckled in the passenger seat as Joanna pressed an older version of the car past the limits of its speedometer. Now she was the most careful of drivers when Sydney was aboard, but Will knew Joanna still occasionally pushed the pedal down when she was alone . . . speeding to escape her past, letting the wind wash her clean.

Will reached for the handle of the driver's door—and stopped himself.

It wasn't that he didn't trust her. There was no doubt in his mind that Joanna loved him fiercely, had in fact never loved any other man.

It was true she had withdrawn from him in the hospital. But he understood to the core of him that she'd believed if she didn't concentrate every cell of her being, keep every waking thought, on Sydney, in that moment Sydney would die. Magical thinking.

Or—the thought came reluctantly—had her concentration, her unwavering focus, actually helped?

Words snaked into his head, unbidden. *Children respond strongly to their parents' beliefs. . . . Your wife may well be giving your daughter the will she needs to turn her own illness around.*

Salk's words.

Will realized with a queasy jolt that he'd almost forgotten the dark man. No, not almost: he'd literally forgotten. He had not once thought of Salk since Flynn's Easter picnic. *How could that be possible?*

Now the memories were as clear as if he were back in the endless maze of Briarwood.

Salk like a specter in the halls; seated too close to Joanna on the garden bench; watching them from the doorway of Sydney's room; tall, dark, seductive.

Through the hot flush of anger, Will recognized an undercurrent to his concern. Jealousy.

Joanna had never lied to Will—never that he knew of—until the moment in the garden when she had denied talking with Salk.

He realized with sudden, sick fear that Joanna had pushed him straight into campaigning, guaranteeing he would be too distracted to notice what she was doing, guaranteeing she would be alone. *To do—what?*

What was she hiding from him now?

He turned to the Cougar, wrestling with himself.

What was worse—not knowing? Or what he was about to do?

He stepped abruptly to the driver's side of the Cougar, opened the door. He fished in his pocket for a business card, stooped, and wrote down the number on the odometer—84011.

He closed the door quietly and stood, conflicted, roiling with guilt.

Then he slipped the card with the mileage into his pocket and turned from the car.

CHAPTER THIRTY-TWO

The sun shone bright on the lagoon of the Public Garden. Swans drifted in lazy circles under the drooping willows. Will stood back on the cobbled path, awkwardly holding a sparkly toy wand with a glittering star on top that they'd bought from a sidewalk vendor at a souvenir kiosk. Ducks clustered quacking at Sydney's feet as she threw bread, ordering imperiously:

"No fighting. Share."

Will watched her, reflected in the rippling water: his mysterious, miraculous daughter. Sometimes just a rambunctious bundle of little girl, sometimes uncannily knowing.

And alive. She's here, alive. Do you really want to open this door?

But his dream taunted him—*Joanna sneaking into the house at dawn* . . . and memories he'd unconsciously or consciously blocked were bubbling up: Sydney waking up, groggy, the night he'd found Joanna in the icy garden:

Daddy, where's the man?

And finding Sydney alone in the hospital room the morning after his long, troubling talk with Salk:

Where's Mommy, sweetie?

She went with the man.

Though he knew beyond doubt that Sydney was not Salk's target, he felt a white hot flash of rage to think of the man with his daughter. How much had she seen, all those days Will had left Joanna alone?

Alone—except for Sydney. The thought was an uneasy twist in his gut. How much did she know?

Will hesitated, looking down at his tiny, perfect, living child. Then he spoke carefully. "Sydney, has Mommy . . . have you seen Mommy . . ."

Sydney looked up at him with his own gray eyes, the sun so bright on her gold hair, the roses in her cheeks.

He couldn't.

"No. Nothing, baby girl."

They continued to walk the path, coming up on the curve of lagoon where Will had first seen Joanna with the swans and his life had changed forever. She was no less ravishing now than she had been that day, irresistible.

He talks to me. . . . He makes me hope.

Will's hands were clenched beside his thighs. He knew that whatever Joanna was hiding now, wherever she was sneaking off to, it involved the dark man.

Beside Will, Sydney slowed, her eyes fixed ahead. She was looking toward the corner of the Common, where a spiked black iron fence surrounded the old Central Burial Ground.

"I want to go there," she said abruptly.

Will was startled from his dark reverie. "Why, honey?"

She shrugged, a child's inscrutability. "Because."

The iron gate opened with a rusty squeal of protest. Sydney walked straight into the graves, and Will followed as she wandered among tall, flat tombstones dating from the Revolutionary era . . . all etched with the same unnerving icon of a skull with wings. The age-old trees with their battered trunks blocked out the sun; it was a good fifteen degrees cooler here than in the park outside the spiked fence. A disquieting place, redolent with history. Will had never cared for it; the aura was too primitive, too reminiscent of Salem, of fanaticism, of hangings and torture, of the country's dark ages of superstition and prejudice.

He trailed behind Sydney, unsettled, as she drifted through the graves, pausing systematically at each of the smaller stones—the graves of children. He stopped behind her to read one:

BELOVED DAUGHTER

Sydney tilted her head, gazing at the stone. And Will's heart stopped as he was struck with a sudden, sick certainty: *She knows. Something's not right, and she knows.*

A shadow passed over the sun . . . darkening the light. Will looked up from the gravestone . . .

Sydney was gone.

He was completely alone.

"Sydney?" His voice quavered as he called her. He took a few halting steps, looked frantically around him at the silent graves, the tall, tangled trees. The sun had moved behind a cloud, and a chilly wind breathed through the uneven rows of tombstones.

It was as if she had vanished into the grave before him. He had a nightmarish thought that he'd just woken up from a wishful dream; that Sydney had died on the operating table the night of her surgery and was buried here beneath his feet.

"Sydney!" He strode through the tilting headstones, his voice bursting out of him now, tearing at his throat. *"Sydney!"* He whirled on the patchy grass, under the trees, every terror of the hospital pressing in on him now, death, guilt, insanity. His gorge rose in his throat . . .

And then Sydney moved out from behind a tall headstone, clutching her wand, looking at him enigmatically.

Will stumbled to her, knelt shakily on the grass, and hugged her for a long moment, breathing in her little-girl smell, his heart pounding crazily, his body still shuddering in slow tremors.

That's what you get for questioning, a voice whispered in his head. *Do you understand now?*

For a moment, a perfectly clear voice, as if someone had spoken just behind him.

But that was madness.

Sydney squirmed in his too-tight embrace. He pulled back, releasing her, and as he stood, shielding his eyes against the sun, he saw a group of children had materialized out of nowhere and were wandering through the headstones like the ghosts of the dead children whose graves Sydney had been examining.

After a startled second he thought, *A school tour, it must be. Just a tour.* But the effect was the familiar feeling of vertigo from the hospital—that nothing was solid or rational.

As he led Sydney out through the black iron gates, the children stood perfectly still among the graves, watching them.

CHAPTER THIRTY-THREE

The sun was just lowering, the trees casting slanting shadows across the lawn, as Will pulled the BMW into the drive, zapped open the garage door.

In the garage Sydney scrambled out of the car, headed for the inner door to the house.

Will got out of the car and stood in the silent, oily darkness, looking toward Joanna's Cougar, scanning for some telltale sign that it had been moved. The car was still parked on the other side of the garage, seemingly in the exact same spot.

Leave it alone, he whispered to himself.

His face tightened. He stepped to the Cougar, pulled open the door quietly, and checked the odometer: 84051.

He knew without looking but took the business card out of his jacket pocket and checked the mileage he'd written before: 84011.

The car had been driven forty miles.

Upstairs, Will stepped through the door of the master bedroom.

Joanna was dozing in bed, an open book beside her, dark hair spilled on the creamy sheets. She seemed not even to breathe . . . perhaps too still.

Will sat carefully on the bed, bent, and kissed her awake.

She moved under him dreamily. "Mmm . . ." She opened her eyes, looked up at him . . . then too quickly away toward the window, the late-afternoon sky.

"Oh, my God. I slept all day."

The lie knifed Will's heart. *Confirmed.*

He kept his face still. After a long moment, he said with effort, "You must have needed it."

Joanna bit her lip, turned her head on the pillow, looking away.

Sydney ran in, brandishing her wand. "Look, Mommy— magic!" She climbed into bed with Joanna, touched the star to Joanna's head. "You're magic, too." Will watched them, mother and child. Joanna glanced at him over Sydney's head—helplessly, he thought—before she turned away.

Locked in his study, door closed to the world, he pored over every detail of their accounts: credit-card bills, bank statements, ATM withdrawals; scouring online and physical files for any evidence of what she was doing, any strange expenditures, travel, meals, anything. After hours of fruitless search, he sank back in his leather chair, staring out the window at the crimson waves of the setting sun over the trees.

He'd found nothing out of the ordinary, no trail whatsoever. The lack of evidence intensified his dread.

He closed his eyes and saw the molten sun still burning against his eyelids. The rage, when it came, was not at Joanna. He clenched his fingers into the armrests of his chair, forced himself to breathe. There would be no assigning of fault, no recrimination. Joanna would never lie except in the most extraordinary of circumstances. He had no doubt that whatever she was doing, whatever she had done, *whatever,* she'd done purely out of love and terror. Salk had offered her something—faith healing, laying on of hands, special intercession with God—something that somehow Joanna had believed—

Will's eyes flew open; he sat forward in sudden alarm. Could Joanna actually be using some unauthorized drug on Sydney?

He forced down the thought. The hospital's constant checkups and bloodwork would have uncovered anything dangerous.

But there was something that Salk had promised and Joanna had believed. And he knew she would have done anything.

Anything.

After dinner, after they'd put Sydney to bed as if it had been any other day, Will followed Joanna into the kitchen to clear the dinner dishes. Joanna poured the last of the Shiraz into goblets, and they drank as Will loaded the dishwasher and Joanna swirled a few pots in sudsy water, the windows steamy behind the row of African violets above the sink. Will watched her; she was, as ever, a mystery.

"Do you ever hear from him?" Will said.

She looked at him. There was just the slightest stop, like a stilled heartbeat, before she answered.

"No. Not at all."

She stooped to scrape leftovers into the cat's bowl, took the dishcloth, and moved into the dining room.

Will stood by the counter. He took a large swallow of dark wine, then stepped into the doorway of the dining room, watching Joanna wipe down the dining room table in the soft candle-light of the centerpiece. After a moment, he spoke.

"I'm dropping out."

She glanced up at him, startled.

"For good this time."

She stammered, uncharacteristically, "But why . . . I don't . . ."

His eyes were on her; he felt his way so carefully. "Things have been spinning. I feel like *we* got lost, somewhere."

Joanna looked down at the table, at the sheen of the wood in the flickering candlelight.

"We've been in hell, Joanna."

He finally moved closer. She looked across the light at him. He spoke from his soul.

"You're my life. There's nothing we can't talk about. There's nothing you could tell me I wouldn't understand." He saw a glimmer of tears in her eyes and felt almost sick with fear. "Whatever it is, Joanna, you have to tell me."

She stood trembling, staring down into the candle flame. After what seemed like an eternity, she spoke from far away. "I did get lost. I think I went mad for a while."

She looked up, her eyes fierce. "But that's over. We're together. I have Sydney. I have you. I could live through anything for that."

She looked at him suddenly, with a rush of feeling.

"Oh, my God . . . Will. Before you, I thought I could never love—or be loved—" Her voice scraped, raw. "I felt filthy . . . and broken . . . not even human. And you came and you loved me and you saved me. You took me out of the dark. If you left me, I couldn't survive it."

Will crossed swiftly to her, took her in his arms. She was shaking, tremors through her entire body.

"Never," he whispered. "Never."

She pulled back and looked at him, her blue eyes black in the candlelight.

"I want you to run. I want you to win. We both want that for you."

He held her, his cheek against her hair, and knew he would kill Salk if he had to.

CHAPTER THIRTY-FOUR

A strangely violent summer storm had hit the coast, drenching the city. Rain pounded on the streets outside Will's Back Bay campaign headquarters, turning the streets into rivers and cascading in sheets down the windows of the storefront office with SULLIVAN NOW posters plastered over every available surface.

Student and senior volunteers wearing campaign buttons and T-shirts turned from their phones to look at Will as he strode dripping by their desks, past the ringing telephones, toward the war room in back.

Jerry looked up from his desk as Will walked in, his suit and hair dark with rain, his face as grim as the sky.

"I want everything you've got on Salk."

Jerry rose. "What—"

"Everything Hank came up with. Who he is, where he came from, where he's gone—"

Jerry stepped to the office door, closed it, muting the phones. He turned to Will cautiously. "Will, the guy doesn't exist."

Will stared hard at Jerry. His friend's face was carefully neutral.

"Hank found nothing. No one at the hospital's ever seen him or heard of a Salk. Staff *or* patients—"

Will felt the familiar wave of vertigo, the sick feeling of reality dissolving around him. He pushed it down, fought for control, for facts. "There are patients who've seen him. I've seen him with patients—"

"Will—"

Will's voice blasted over Jerry. "He's been in contact with Joanna—"

"No, he hasn't," Jerry said sharply.

Will stopped, arrested by the certainty in Jerry's voice.

Jerry sighed, sat back in the swivel chair. "I've had her followed." He continued quickly, "Will, I had to—there's too much at stake."

Will said, dry-mouthed, "Just tell me."

"She's going to a neighborhood in the South Side. Wrong side of the tracks." Jerry reached into a bottom drawer, pulled out a file. "She goes to this house."

He opened the file to reveal photos of a half-burned-out house with a sagging porch, leprous paint. Interior photos showed furniture broken and charred in some squatter's attempt at a campfire, ransacked cabinets, a rusted iron bed with a filthy mattress. Will stared down at the images in complete incomprehension.

"It's vacant. No one's lived there for years. We thought . . . you know, she was going there to score." Jerry finished in a rush before Will could explode, "Come on, Will, a woman like Joanna in that neighborhood? What else would she be doing? And who could blame her? Jesus—after all you two have been through?" Jerry's voice was forced, reassuring. "But Hank hasn't seen or found anything. No vials, rigs, nothing. As far as we know, she just goes inside and sits. By herself. An hour, sometimes two. No one else comes or goes. After a couple of hours, she comes out and goes

straight home. And we figured—you'd have seen . . . something."
There was an implied question in Jerry's pause.

"No," Will said, dazed. His mind was racing through the possibilities. Could he have missed something this major? Was
Joanna's withdrawal from him in the hospital partly drug related?

His mind seized on the next logical thought. Was Salk supply-
ing her? Was that the power he had over her?

"Who owns the house?" he demanded, his throat tight.

"The city," Jerry responded promptly, handing over a photo-
copied deed from the file. "The whole block's been condemned.
It's slated for a tear-down, but . . ." His shrug said, "Not anytime
soon." Governor Benullo had shown little interest in improving
any infrastructure he couldn't slap his own name on.

Jerry was watching Will warily. Will snapped at him, "And
who owned it before that?"

"An old couple held the title, rented it out. They died within
three months of each other, ten years ago. House hasn't been oc-
cupied since." Jerry motioned with a finger to the second page of
the document Will was holding. Will flipped the page to look
down at another deed, an attached typed list of names that meant
nothing to him.

He dropped the papers on Jerry's desk. "This is about Salk.
Find him. Get Hank back to the hospital. Any connection you can
find—"

Jerry stood. *"Will."* His voice was so harsh, Will stopped. Jerry
held up a clenched hand—forced himself to open his fingers and
spoke slowly, with emphasis. "You have got to let this go. The hospital was the hospital—I know what kind of strain you were under. But you were *losing it.*"

Will stepped back, silenced. It was the one point he was unable
to argue. Jerry sighed, sat, pressing his hands into the desk, holding Will's eyes.

"You have to look after your family now. Whatever Joanna's

doing, she's sure as hell not meeting anyone. But it doesn't look good, Will. She's been careful so far . . ." He hesitated, then repeated, "It doesn't look good. The national spotlight is on you. You could lose everything."

Will looked at him sharply, suddenly alert to a possible ulterior motive.

Jerry didn't seem to notice. He went on, soothing, "Why don't you take a few days off, take Joanna to Maui or something. Okay? Get out of here, get some sun. I'm sure whatever this is, you can talk it out, and once you have the facts . . ." He paused diplomatically. "If she needs to see someone, go someplace for a week or two, we can make that happen. No one needs to know."

Will slammed the BMW door closed and wiped the rain from his face.

He sat back against the leather seat, staring out the drenched windshield, his mind going a million miles a minute.

Drugs. Salk. That foul house.

Could it be?

Joanna had been driven to the depths of desperation. Will had no doubt. Nothing was impossible.

But for all Jerry's documentation, something was way off. The most basic detective work could have turned Salk up. Will was loath to believe Jerry was lying to him, but the possibility was unmistakable. At the very least, he was withholding information to protect the campaign.

He'd succeeded only in making Will more convinced that Salk was a threat.

Will pulled his hands from the steering wheel, turned the key in the ignition. Whatever this was, he knew where he had to start.

CHAPTER THIRTY-FIVE

The windshield wipers beat against the steady rain. Will drank bitter black coffee, watched from where he was parked on the boat pad, shielded from his house by a row of trees. He slid down on the seat, hunching like a PI from a bad TV movie as Joanna came out of the house onto the porch with Sydney. They stood still for a moment, looking out at the rain, matching silk-screened umbrellas in hand. Joanna looked up at the sky, and Sydney looked up with her, a tiny, perfect mirror.

Joanna was dressed in a chic, slim dress and sheer stockings, which instantly churned Will's stomach.

She was never less than heartbreakingly beautiful, but this seemed more calculated than her usual careful grooming: the way the plum silk (a color that emphasized her lush darkness) hugged her slim waist, outlined her breasts. Will clenched his hands on the steering wheel taunted by the picture of the burned-out house, imagining Joanna on the filthy bed. He fought for control as mindless jealousy, thoughts he didn't want to think, washed over him in waves.

Joanna took Sydney's hand and led her back in the house,

closing the door behind them. Will sat back in the seat and tried to breathe, watching.

After a moment the garage door rose, and the black Cougar drove out of the garage.

The rain turned to dense fog as Will followed the Cougar over the harbor bridge into Boston, staring ahead of him at the faint red taillights through the mist. All his instincts told him that whatever Joanna was doing in the abandoned house, she would not bring Sydney with her to witness it. And as he'd thought, the Cougar passed the turnoff to the South Side that would have taken them to the address Jerry had given him.

Will knew instantly where she was going; it was as inevitable as breathing. She exited at the Briarwood Avenue off-ramp and turned toward the hospital. He followed her, crawling through traffic thanks to a three-car collision in an intersection up ahead. Bystanders gawked from the sidewalks in front of cafés and food marts; sirens screamed as ambulances and tow trucks arrived from all directions. Will tensed at the too familiar sound and turned up the Bach on the stereo to try to drown it out. Ahead, beyond the overturned cars, the hospital loomed against the clouds like a massive animal, crouched, waiting.

Finally the intersection was cleared. Will held back in the BMW as the Cougar turned off into the long circular driveway of Children's and drove down into the dark mouth of the underground parking lot.

Will braced himself . . . and followed.

He stepped off the elevator from the garage into the main lobby of Children's, into the cheery blast of multicolored neon lights, the aquariums and dioramas and gift kiosks and the house-size book with its endless list of names of hospital donors.

Will turned to the escalator and fled the falsely carnival atmosphere.

On the second floor, he was enveloped immediately by sense memories of panic and disorientation, the nightmare fog of the hospital. The pulse was there around him, ever-present—the blip of monitors and the whispering of patients' families. Parents walked past him in an all-too-familiar daze, leaning into each other for support; wizened children watched from doorways with hollow, sunken eyes. The smells seemed heightened, nauseating—the sharp bite of alcohol and the sickly-sweet stench of chronic illness.

Will flinched away and walked faster through the gleaming corridors. He had not been back to Briarwood since Sydney was discharged, and the realization pricked him with guilt. It was Joanna, always Joanna, who took Sydney for her weekly checkups and bloodwork, the biweekly MRIs. It was Joanna who gave the glowing reports of Sydney's progress. Will had simply listened and accepted.

The elevator up to the cancer ward was a level of unease all its own. Will stood stiffly, trying not to think about the inner halls concealed just behind his back, like termite tunnels in seemingly solid walls. But his heart was pounding and his palms sweating by the time the doors breathed open and he escaped into the yellow lobby of the Children's cancer ward.

He walked up to the circular nurses' station. Sandra, one of the regular receptionists, smiled up at him.

"Hi, Mr. Sullivan. How nice to see you. How's Sydney?"

Will felt a rush of dread but kept his voice calm. "Aren't they here today?"

Sandra looked puzzled, scrolled through the appointment chart on the computer, polished nails clicking on the keys. "I don't have her scheduled for anything."

Will was barely able to smile a good-bye as he turned away and into the maze of yellow corridors. Operating purely on instinct, he rode the elevator down to ground level and crossed the cool green inner hall to the wall of glass to look out into the children's side of the garden. Absent of snow, the garden was startlingly luxuriant—lush, exotic plants and flowers in full bloom.

The rain had relented, leaving water droplets on the glistening leaves and a thick ground mist, adding to the primeval illusion. Joanna was nowhere in sight. But in the sea of mist, he spotted Sydney playing with some child patients among whimsical toadstool chairs, statues of gnomes and fairies. The children were engaged in some stalking game, moving through the exotic plants with croquet mallets, as if in a dangerous jungle on safari. Sydney suddenly lunged and beat at an invisible beast with her mallet. The other children followed suit, raising their mallets and beating the invisible creature to death.

Will felt gooseflesh rising at the uncanny violence.

There was movement behind him, and he spun.

Joanna stood in the dim greenish hall, looking surprised but guilelessly pleased to see him.

"That was a fast luncheon."

"Canceled," he said shortly. "I didn't think Sydney had an appointment today." He knew he sounded like a prosecutor. She raised her eyebrows slightly.

"She doesn't. We were going shopping, and I thought I'd stop and pick up this month's prescriptions." She lifted the pharmacy bag in her hand—evidence. When Will didn't speak, she tilted her head, puzzled. "Is something wrong?"

He spoke through a dry throat. "I don't know. Is there?"

Again, she looked faintly perplexed. "What could be wrong?" She stepped up to kiss him; even that light brush of her lips was an erotic charge. "Any chance of you taking us to lunch, then?"

He shook his head, feigning annoyed regret. "Some Kiwanis thing at two."

"Lucky them. Well . . ." She ran a hand up his shirt, straightened his tie. "Be good."

He looked into her eyes. She met his with a clear gaze before she kissed him lightly again.

Not a tremor. Perfectly normal. Except that she hadn't asked him what *he* was doing at the hospital.

Will descended the escalator toward the neon-lit lobby, his thoughts in turmoil.

And what now? Follow her further? Where would she really go, with Sydney in tow?

Outside, the rain had started again, pouring down the windows in waves. The ten-foot-long aquarium and blue neon added to the underwater illusion.

Will started for the pneumatic doors, then slowed, looking across the lobby at the house-size book of hospital donors. On impulse, he walked past the aquarium over to the display and stood in front of the enormous open book, scanning the lists. In the S's, he found his own name and Joanna's, his mother's and father's—but no Salk.

Disappointment was bitter in his mouth, but it had been a long shot. He started to turn away . . . and saw a glass case of photos on the opposite wall, a gallery of hospital history. Frowning, he moved over to it, walking slowly by the collection of photos: sepia-toned aerial shots of the original buildings, vaguely sinister black-and-white photos of stiffly posed uniformed nurses, intent bespectacled doctors in primitive-looking labs, physical therapy rooms with patients immersed in enormous vats.

Will moved past with an indefinable feeling of unease. In more modern color photos, again, his own family was prominent: Hal and Susan at a groundbreaking ceremony; at the dedication of a new transplant wing, with Senator Flynn and his wife beside them, an eight-year-old Will holding his mother's hand.

There were other governors in the cavalcade as well. Will grimaced at a photo of Benullo, smirking as he raised a shovel to scoop a first load of dirt—

Will froze.

In the photo, behind Benullo, Salk stood with a group of suited goons, looking on.

Will jolted forward, staring through the glass. The dark man in the photo was clearly Salk—exactly as Will had last seen him: the same elegant carriage and suit; the quiet, intense focus.

On Benullo.

Of course. Of course.

Along with the rage, Will felt the lightness of relief.

He glanced around the lobby, saw children playing on the jungle gym . . . distracted, waiting families . . . the secretaries at the round reception desk.

He reached for the sliding glass of the display case, pulled on it. Locked, of course. But a cheap, standard cabinet lock, and Will had worked too long with too many cops not to have picked up some elementary skills.

He found a paper clip on the gift shop counter and twisted it straight in his fingers as he circled casually back to the display case, his coat draped over his arm.

Five minutes later, he escaped out the pneumatic doors of the entrance, a sheen of sweat on his face—and the photograph in his pocket.

Wind whipped rain between the buildings as he walked stiff-legged down the marble stairs, oblivious to the downpour. The photo felt on fire in the lining of his coat. *Proof. Finally, some proof.*

The pedestrians on the sidewalks and access streets around him were almost entirely medical personnel, in white coats or blue scrubs covered with clear plastic raincoats, wheeling IV stands, pushing carts of buckets labeled BIOHAZARD.

Will was rigid with tension as he walked, rain soaking his hair, blurring his vision. His mind was racing; he walked with no destination, only the explosive need to move. So it *was* Benullo, all along. Hiring Salk, most likely. Sending him after Joanna, because Benullo, scum that he was, had a primitive instinct for weakness and would have sensed Joanna's vulnerability. And would delight in engineering Will's downfall through her. Will's whole body was burning with rage, the desire to annihilate the man.

He stopped under the eaves and pulled the photograph from his inside jacket pocket to examine it. The date in the bottom corner of the photo was the first year of Benullo's mayorship, the same year Benullo's wife had died unexpectedly—

At Briarwood, Will remembered with a start. She'd been a patient at the Fordham Clinic. Will's mind scrambled for the exact details. She'd been struck with some kind of debilitating muscle disease shortly after Benullo was elected and died just weeks afterward.

The thought was disturbing on a level Will couldn't quite fathom. He pocketed the photo, suddenly loath to touch it.

As he resumed walking, he became aware that the echo of his footsteps was arrhythmic, off a beat. He slowed on the slick stones, listening—and was certain. Someone was following him, trying to match his footsteps.

Will turned abruptly into the courtyard separating ultramodern Doctor's Hospital from the marble buildings of the old Mass Bay Medical College. He strained toward the sound as he moved through puddles on the walkway beside the landscaped green between the buildings. The footsteps were still following.

Will casually shifted his direction, heading diagonally up the wide steps of the next hospital. Out of the corner of his eye, he caught a glimpse of a tall man in a dark overcoat.

Will felt himself flush with anger. He moved onto the pillared walkway at the top of the steps and walked faster, weaving through the columns. The footsteps moved faster behind him, the sound echoing off the marble facade . . .

Will turned the corner at the end of the building . . .

. . . and stopped, pressing his back against the granite wall, waiting, watching the corner. Rain pounded on the walkway . . . the footsteps approached . . .

I've got you now, you bastard.

The tall man in the overcoat dodged around the corner—

Will grabbed his pursuer, slammed him against the wall. The man held himself stiff but unmoving in Will's grasp. "Whoa, pal. It's cool," he said quietly, and through his fury Will finally recognized Hank, the PI he'd worked with for years in the DA's office.

Will released him slowly. "What the hell are you doing?"

But he knew before he'd finished the sentence. Of course, Jerry was having him followed, too.

They sat on a bench between the marble columns, the lean, hawk-eyed private investigator fidgeting in the speedy way Will remembered—a holdover from too many years as an undercover drug agent, barely conquered demons of his own. The rain was a curtain between them and the plaza, shutting out the rest of the world.

"I need to know everything," Will said without further explanation.

Hank shifted abruptly, his leg bouncing on his knee. His accent was as broad as South Boston. "Jerry said he gave you the report."

"I need to hear it from you."

"Will, I got nothing. The guy doesn't exist."

The rage rose in Will again. "That's bullshit. There are patients who've seen him—"

"Then no one was talking. No one. *Will*," Hank said, injured. "What'd you want me to say? You know I don't shit around with this stuff."

Will reached into his coat pocket, pulled out the hospital photo, and passed it to Hank. He watched the PI's face as he looked down at the images, saw the jolt of surprise.

"The tall one behind. That's him," Will said quietly.

"Benullo. Holy Christ." Hank stared down, his face working, then looked up from the photo, shaken. "Will, I'm sorry, I— Jeez. I really didn't—I couldn't find anything—"

Will spoke levelly. "The guy's good. No doubt." His face hardened as he turned into a prosecutor again. He stabbed a finger at Salk's image. "You've never seen him before?"

"No."

"Never with Joanna?"

The sheen of wet on Hank's face was sweat, not rain. "No, Will. No. I swear."

Will exhaled, contained himself. "Tell me about the house. You've seen her there. On Cabarrus Street."

Hank nodded assent.

"How often?"

"Pretty much every other day. And a couple of times at . . ." The PI paused, clearly reluctant to go further.

Will fought to keep the shock from showing on his face. "At night?" he asked tightly.

The PI nodded . . . and Will had to look away from the quick pity in his eyes. He stared out over the rain-drenched plaza.

"No one else comes? No one?" Will demanded.

"Will." Hank's voice was full of reproach.

Will sat back on the marble bench, pointed at the photo. "Find him." Hank started to nod agreement. Will held up a hand. "And this isn't for Jerry this time."

Hank's leg bounced up and down on his knee, and Will wondered for a moment if he was on something. "Sure, Will. Sure. You got it." He slipped the photo into his coat.

But as they stood, a maddening voice in Will's head whispered, *Hank's good, but Salk is better. Not even the same league.*

He drove home in the steady fall of rain, piecing together how it must have gone, remembering with hot shame: Joanna turning from the bed table, offering him a glass of champagne . . . pushing him slowly back on the bed, bending to kiss him, and gushing the wine into his open mouth . . . and the deep sleep he'd chalked up to his campaign schedule, to the many months of exhaustion.

She'd drugged him. She must have. So that she could do—what?

Heat flushed his body; he pounded the dashboard with one hand, needing the pain.

Hank had sworn he'd never seen Salk with Joanna, but Salk had managed to cover his tracks in every other aspect of his being, so why not with Joanna? If Benullo had hired Salk—or if Salk was part of his team—then it was clear they were trying to get to Will through Joanna, to create a scandal that Will's campaign, and much more important, his marriage, could not survive. Will's blood boiled at the thought.

How far would Benullo go?

He knew the answer instantly: As far as he had to. The governor loathed Will and would take a singularly prurient pleasure in corrupting Joanna.

Home again, Will slammed into the house. The entry was dim, the rooms silent and empty. Will's pulse skyrocketed as something brushed up against his shin . . .

Only the cat, snaking around his shoes. He stepped over it and strode for the stairs.

In their bedroom, Will avoided looking at the bed as he crossed to the bathroom, where he systematically ransacked the cabinets, searching through the rows of Sydney's medications, hypodermics, which he examined, frowning.

Then he reached abruptly to the back of the cabinet and removed an almost empty bottle of Vicodin. He checked the label, calculating. Refilled a month ago, and just two pills left out of a prescription of a dozen.

But Sydney hadn't had any kind of pain at all since the surgery.

And why was that? Surely not normal.

But he pushed that thought away.

That night, he poured out the wine Joanna gave him.

And waited.

CHAPTER THIRTY-SIX

The fire had burned down to red coals. Will lay still in the softness of their sheets, feeling Joanna awake and alert beside him, her body as tense as a thief's. Without breathing, she lifted the duvet, eased herself out of the bed . . . and backed into the shadows, silent as the night.

He waited until he heard the creak of her step on the stairs, then sat up in the dark and punched in a number he'd programmed into his cell phone earlier in the day. He dressed at the window, watching Joanna back the Cougar out of the garage without lights, realizing in some part of his mind that the crackling of the lit bedroom fire would have masked the minimal car noise the times she had left before. Still, she must have been using several pills at a time to keep him from waking.

He went downstairs without turning on the lights and opened the door to Valerie, a campaign volunteer he'd enlisted to stay in the house with Sydney, feigning an early-morning business trip. Will pointed upward, spoke low. "Fast asleep."

Valerie settled herself in the living room with an economics textbook, and Will was on the road, his own headlights the only moving thing in the darkness of the woods.

He'd taken a chance, letting Joanna start off without him, but was gambling that Hank and Jerry were correct about her destination and that he could make up the time on the road. It was all too strange not to be true.

He sped on the deserted turnpike, across the bridge toward the hospital. Rain had turned to thick fog; the mist on the bridge felt nightmarish, hovering over the water, deadening sense and sound. Will was grateful for the adrenaline keeping him awake and focused. At one point he reached for the radio, but the sudden blast of music was unbearable and he punched it off again. He continued, across the railroad tracks, over water from the day's downpour, pooled in the gravel trenches.

The off-ramp he'd MapQuested descended into an industrial area of town. Will stared ahead through the dark windshield with growing dread. The neighborhood he was headed into had never been good; now it was in complete decay. Boarded-up buildings, a liquor store, an auto parts junkyard behind barbed wire.

Homeless people camped on the streets, shopping carts full of belongings beside them. Through the curling mist, Will saw the lights of trash can fires down alleyways; the few furtive loiterers had the hazy eyes of crack addicts.

His gut twisted, that Joanna was exposing herself to this neighborhood for even one night. *Could it be about drugs?* He stared bleakly around him at the walking dead, peering out from doorways. *How could it not be?*

He drove slowly to the corner. The street sign read: CABARRUS STREET. He made the turn with apprehension.

Halfway down the short block of houses, Joanna's Cougar was parked without lights in front of the derelict house from Jerry's photograph: half-burned, with windows broken, porch sagging and splintered. Ancient CONDEMNED signs were plastered on the windows and door. There were no other functioning cars anywhere near.

Will parked down the street behind a Dumpster at the curb and shut off the engine. Silence surrounded him; mist fogged the view out the windshield. He glanced at the clock on the dash: 2:11 A.M. He clenched his hands on the steering wheel, then reached for the glove box and withdrew a police-issue nine-millimeter Glock.

Of course he had one. No city prosecutor was naive to the constant threat of retaliation by the bad guys. Will didn't often carry his weapon, but he'd learned long ago how to use it, and well. It had never been his own safety that concerned him.

The gun felt heavy and lethal in his hand.

He got out of the car and stood on the cracked sidewalk, looking up through the mist at the house. It was vile, dark, crouched balefully on the decaying street. Will felt a wave of revulsion at the thought of Joanna inside the loathsome place. The hair was standing up on his arms; the sense of not-rightness was overwhelming.

His grip on the gun was so tight, his hand cramped and he had to force his fingers loose, his palm dripping with sweat. He knew he was on a dangerous edge: If Salk appeared, he would kill him without dialogue, without pause for explanation, without thought for his life or career—he would shoot until he had eradicated the essence of the man. Without question.

He exhaled, moved quietly across the weed-choked patch of yard to the stoop, and hovered, looking up at the crumbling porch, listening.

No sound from within.

Gripping the gun, he moved around the side of the house, avoiding piles of trash and broken bottles. He stepped cautiously up to look in through a filthy windowpane.

Through the smeared glass he could see a small, dim room with burned-out walls, a metal-frame bed in the corner with a stained mattress.

His breath caught in his throat.

Joanna was sitting on the bed. Will's eyes swept the shadows . . . but she was alone in the room. She sat without moving on the rotting mattress in her expensive dress, her arms limp by her sides, her eyes blank.

Will stared through the window at her face, his heart pounding with dread and confusion. She was so still, her features slack, barely present. Had she done drugs before he'd gotten there— shot up, smoked crack in the space of time that he was following her? He scanned the bed, the floor in front of her. There was no pipe, no rig.

Will clenched his fingers around the gun, fighting a powerful urge to break the window, to wrest her out of the place. But he had to see for himself. If Salk was coming, he had to know.

He kept his vigil for over two hours, rooted to the hard-packed dirt beside the house, the fog moist on his skin, the acrid smell of industrial waste in his nostrils. The window was in his sight at all times. Inside the burned room, Joanna sat without moving. Once in a great while, a dilapidated car passed without stopping. No one even came close to the house; no one entered—and no one left.

Finally, through the window, Will saw Joanna rise from the bed with no more consciousness than she had demonstrated the entire time he had been watching.

Will backed up, ran noiselessly across the scraggly yard to his car, slipped inside, and slid beneath the steering wheel just as the front door opened and Joanna came out. Gasping in shallow breaths, he peered up through the rim of the steering wheel, out the windshield, but there was no reason for caution; Joanna seemed to see nothing. He strained for a glimpse of her face, and for a moment she was caught in the moonlight. Her eyes were dead; there were tracks of mascara on her cheeks from crying.

She listlessly opened the door of the Cougar and dropped inside.

Speeding on back roads to beat her home, Will was in bed, baby-sitter dispatched with another lie of a canceled flight, when Joanna crept back into the bedroom. He made himself breathe regularly as she undressed in the bathroom and laid herself carefully down beside him.

He could feel the night on her—cold.

CHAPTER THIRTY-SEVEN

He was slow to surface from sleep; his limbs felt heavy, his head numb. He turned over, burying his head in a pillow, reluctant to open his eyes, to be fully awake. Lying motionless in bed, he could hear Joanna and Sydney downstairs in the kitchen, could smell coffee brewing and bacon sizzling. The strange journey of the night could almost have been the lingering vexation of a vague dream.

Let it be a dream.

He came down the stairs, dressed for business, and stepped into the kitchen doorway.

The scene was perfectly normal: Joanna frying bacon at the stove; Sydney drawing with crayons at the table. She looked up, beaming, as Will moved into the kitchen.

"Daddy, I made flowers!"

Will moved over to look at Sydney's picture—a happy garble of colors.

"That's beautiful, Princess." He had a momentary flash of Sydney in the hospital garden, savagely beating the invisible creature to death, and pushed down the thought.

His eyes went to Joanna. She had just lifted the skillet from the stove, but her wrist remained in the blue flame of the gas burner. The flesh was searing. She didn't pull her arm away.

Will jolted toward her. *"Joanna!"*

She started as if coming awake, jerked her arm away from the fire.

Will was at her side in a second, taking the skillet from her, guiding her to the sink. "Jesus . . ." He ran cold water over her wrist; the skin blistered under the stream.

Sydney stood on her chair, craning to see. *"Ouch,* Mommy!"

Joanna stammered, "How stupid . . . I didn't notice . . ."

Not notice her arm was burning? Will looked at her. She was so pale. . . .

The phone started to ring behind them.

Will looked down at Joanna, at her scorched arm. She forced a smile, but her eyes were clouded with pain.

"Go on—get that. I'm fine, it's just a burn."

Sydney scrambled down from her chair and took Joanna by the hand, tugging her toward the hall.

"Put on the slimy stuff."

Will watched them go . . . finally stepped to the ringing phone. He picked it up—and punched it off.

CHAPTER THIRTY-EIGHT

Shadows from fast-moving clouds cast patches of light and dark over the hospital buildings. The vast complex loomed as Will drove into the underground parking lot of Children's.

He sat in the car in the dark parking lot, seeing Joanna with her arm in the flame. He could no longer rationalize what was in front of his eyes: her night journey to the decaying neighborhood, the secretiveness, the stupor in which she'd sat, her delayed reaction to pain this morning. Drugs were the only plausible explanation. And Jerry was right: Considering the prolonged hopelessness she'd endured in the hospital, the stress of caregiving, anyone could have broken. Anyone.

Will's jaw tightened; he tipped his head back against the head-rest. The hospital photo proved Salk was connected to Benullo, had been connected to him for years. Will was more and more clear about how it must have gone: When Joanna was losing her mind in the hospital, Salk had won her trust, then given her drugs, promising relief. Will couldn't deny the crude effectiveness of the plan: setting Joanna up to be exposed for drug use in order to torpedo Will's candidacy.

Will fought the urge to storm the State House, to confront Benullo right there in his own office. He knew letting Hank handle that investigation was the smarter play. In the meantime, there was evidence he could find on his own. Hank hadn't been able to track Salk from the hospital—Will could.

But there was something he was losing in this confusing flood of new information, a puzzle piece he was leaving out. A thought prodded the back of his mind, then was quickly gone. *Find Salk and the rest will follow.*

He reached for the door handle and got out of the car.

In the familiar yellow upstairs lobby, Cass sat in the curve of the nurses' station filling out a duty roster, her hospital smock a riot of tropical colors. Will stood in the entry of the lobby, watching her; she seemed to be in a circle of light on the dark day. He moved forward, up to the desk. She looked up, and her face brightened. She started in her hearty voice:

"Well now, it's about time you came to . . ."

But she trailed off, seeing the tension in Will's face. Something flickered in her eyes.

"What is it, Mr. S.?"

Will spoke without preface. "Cass, I'm investigating what I think may be a . . ." He paused and then said the only rational thing he could say. "Con man—working the hospital. I think he's running some kind of" —again he hesitated—"extortion scam on some patients."

A shadow passed over Cass's face. "A con man."

Will ignored the shadow. "Yes. I need to find him."

"That man you asked about before?"

Will seized on the uneasy ambiguity in her voice. "Are you sure you haven't seen him? Or heard anyone else talk about him? He calls himself Salk. . . ."

The big nurse was watching him; her probing gaze seemed to see more about him than he knew himself. She spoke slowly. "All kinds get in here, Mr. S. The hospital attracts them. All these people in pain . . . a lot of low-downs think they can take advantage."

Will pressed her. "I need contact information for some of the patients he's approached. Sarah Tennyson. An Officer Sean O'Keefe, BPD. A patient named Gary—he would be in the AIDS ward—"

Cass was shaking her head. "Can't get to patient records if they're in another hospital."

"For Sarah Tennyson, then."

Cass frowned but typed into the computer, copied an address and phone number from the screen onto a memo pad. Watching her, Will had a sudden memory of standing with Salk in the doorway of Anthony Tennyson's room while Sarah and her little boy packed to go home. He spoke abruptly, unaware of what he was going to say until it was out.

"Anthony Tennyson went into remission suddenly, a few days before Sydney." He looked across the counter at Cass. "Did you think there was anything strange about that case?"

Cass's face changed, all hesitation gone. "Childhood leukemia is curable in eighty percent of cases, if you catch it quick," she said with pride.

Will's eyes clouded as again something tugged uneasily at his mind. "But Sydney's type of cancer . . . doesn't have the same recovery rate," he said slowly.

Cass dropped her gaze. "Mr. S., you know you ought to be talking to a doctor about that."

There. It was unmistakable, the deep unease in her voice. About Sydney. Will felt a chill in the too warm room.

"I want to talk to a doctor," he said with sudden force. "I want to talk to Dr. Connor." He mentally kicked himself for not having thought of the young doctor sooner. Another missing piece of the puzzle. "I was told he took a residency in another hospital."

The last sentence was a subtle question. Cass understood. She looked out at the rain. "Yeah, I heard that," she said finally.

"I need to find him, Cass. Can you get some contact information for me?"

She looked troubled but nodded once. "Okay, Mr. S. I'll do that."

He reached across the counter and touched her hand as he took the memo with Sarah Tennyson's address. "Cass, if there's something you know . . ."

Again, there was that flash in her eyes. "What I know is you been through bad times. Maybe not over yet. You got questions, maybe you don't need to be asking me."

Will walked in the corridors, irked by the hint of fundamentalism in Cass's rebuke. Without noticing, he had turned the corner toward Sydney's former room, and he slowed, feeling the old sensation of dread and hopelessness. The soothing pastel yellow of the walls around him seemed sickly, diseased.

A door to the left ahead led out to the balcony. *The balcony.* Will stopped and stepped out the door into a strong, cold wind that tore at his coat, stung his cheeks.

He shoved his hands deep into his pockets and looked to the empty railing. Salk had certainly been here, standing at the rail, as real as Will was. Sarah Tennyson had talked with him, and Officer O'Keefe's wife, and Gary, even if his lover had been unaware of the visits.

But even as he reviewed the names in his mind, Will was conscious of a disconnect. Why would Salk bother with any of these people if his target was Will, through Joanna? What could any of them have to do with Benullo?

Unless the encounters Will had seen between Salk and the other patients had merely been staged, for Will's benefit.

But why?

He looked over the hospital complex: the curves of Children's; the white marble Doric buildings and courtyards of the old Mass Bay Medical College; gloomy, Gothic Mercy.

He frowned and looked harder, vaguely aware that something wasn't right. Dark clouds scudded over the tops of the buildings, creating the unnerving illusion of the hospital growing, looming up. Then Will's heart plummeted.

There was no sign of the dome of the rotunda.

Back inside the maze of yellow corridors, Will headed for the nearest information desk. It was staffed, as most of them were, by an eager senior citizen volunteer, angels of the hospital whose hearts were undeniably in the right place but who rarely had any useful information to impart.

The pink-sweatered grandmother beamed up at him, then frowned in vague puzzlement at his query. "The rotunda? What building is that in, dear?"

"It's between buildings, I think," Will said slowly, trying to re-call how he'd found the place to begin with. "A huge dome, with a—hand . . ." He felt the strangeness of the words even as he said them.

She was already turned away from him, her hands fluttering over the desk, reaching first for the phone, then for the spiral-bound hospital directory. "Let me call . . . let's see now, who would know?" She opened the booklet at random, felt for the glasses hanging around her neck.

Will leaned in and touched her wrist lightly, gave her his warmest smile. "You know, I remember now. Thanks so much for your help."

He turned away, his smile dying, and walked for the elevators, his whole body tense. He knew where he had to go.

Just as he was reaching for the elevator call button, his phone buzzed in his jacket. He checked the caller ID, clicked on. "Hank."

"Will. I got something."

Will felt a shock of adrenaline. "Tell me."

The PI's voice was speedy, and frightened. "Will, you were right. Your guy and Benullo. You were right. Jesus. It was the *wife*—"

Hank's voice cut off abruptly. Will spoke sharply into the phone. "Hank. . . . *Hank*—"

There was a long silence, then the PI's voice came back on, low and hoarse. "Not on the phone. Meet me at the house. You know." The voice paused. "Eight tonight."

CHAPTER THIRTY-NINE

The sky had opened again, cracks of lightning forking through the night. Rain battered the car as Will made the journey over the bridge and mist-shrouded bay, darker than ever in the pouring rain, across the railroad tracks, into the desolate urban wasteland. The streets were eerily empty. There were no homeless in the alleys tonight; the storm had forced them into the abandoned buildings. The sense of déjà vu was strong as Will passed the auto parts yard with its barbed-wire spirals, made the turn onto Cabarrus Street, past the row of scraggly trees. This time it was Hank's Impala parked in front of the decaying house, rain bouncing like hail off the roof and hood. Will stopped again behind the Dumpster and turned off the frantically beating windshield wipers. Again he reached for the nine-millimeter in the glove box. Every repetition of last night's movements intensified his sense of foreboding.

He got out of the car and ducked against the deluge, splashed through the water-swollen gutter, and hustled up the sidewalk toward the sagging house with the CONDEMNED signs.

On the splintering planks of the porch, he shook water off his coat, wiped his face with a hand. He faced the door and felt a visceral reluctance to enter the house, a queasy clutch in his stomach. He put his hand to the doorknob and turned it; the cold knob twisted easily in his fingers.

Will stepped over the threshold and stood in the narrow, dank hall, letting his eyes adjust to the dark, cursing himself for not bringing a flashlight. The hall split the house in half—a living room and bedroom on one side; the burned remnants of a kitchen, bath, and bedroom on the other. Water poured through holes in the roof, spread in pools on the uneven floorboards. Beer bottles and potato-chip bags and other filth lay moldering in corners; there was a rank smell to the house, smoke and urine and something else. Will tensed; his flesh was crawling. He could hear no one in the house. Some instinct made him not call out Hank's name. Instead, he drew the Glock from his pocket.

Stiff with misgiving, every sense on alert, he moved noiselessly down the hall. The stench grew stronger as he approached the last door, the room he'd seen Joanna in.

A sudden scuttling sound raised the hair at the back of his neck. He whirled, lifted the gun—and saw the red eyes of a possum-size rat glaring balefully from the other side of the hall, whiskers twitching. Will swallowed against bile as the rodent retreated sluggishly into the kitchen.

He turned and stepped through the last door, into the small, dim room with burned-out walls.

Hank was slumped on the metal-frame bed in the corner, his head dropped back against the wall.

Will lunged to the bed and clamped his hand on Hank's neck, already knowing it was too late. The PI's eyes stared toward the ceiling, open and dilated; his skin was gray and cold to the touch, his limbs going stiff. A piece of rubber tubing was wrapped around one bicep; there were two syringes and a bent spoon on the rotting mattress beside him. Old track marks traversed his arm.

Will stepped back, breathing shallowly, forcing himself to think through the adrenaline rush of fury and fear.

Not an accident.

Salk? Benullo? Salk working for Benullo?

If Benullo could have Hank killed, there was nothing he wouldn't do. Will's mind was racing. *Had he walked into a trap?*

He dug his cell phone from his pants pocket and stepped to the doorway to keep watch as he punched in 911. He spoke shortly to the dispatcher. "This is District Attorney Sullivan. I need an ambulance and a homicide unit at 95 Cabarrus Street."

Will punched off and immediately speed-dialed Jerry, moving out of the room into the dark hall to avoid looking at Hank's corpse. There was a click on the phone as Jerry picked up, and Will spoke in the darkness without preamble. "Hank's dead."

He heard Jerry draw in a breath. "Jesus, Will."

"Overdose. He's at the house on Cabarrus Street. He called me to meet him, and I found him. Cops are on their way."

For a moment, Jerry's silence was more stunned than calculating. "He wanted you to meet him why?"

"I don't know," Will lied. "He didn't tell you?"

"No." Jerry was quiet again, processing. Will stood motionless in the dark, breathing the fetid air, waiting. "All right," Jerry said. "Tell the cops he was working for the campaign. Security—you don't know what, specifically. I'll handle the rest of it when they talk to me. Say as little as possible." His voice suddenly changed, wary. "Is Joanna there?"

Will's heart lurched. "No. No."

Jerry exhaled. "Good. Call me, then."

Will punched off, reeling from Jerry's question, the insinuation there. Sirens wailed faintly outside, approaching. He stepped into the room again, moved to the bed, and quickly, carefully, searched Hank's pockets. The PI had no wallet, no weapon, no documents on him, and of course, no photograph of Salk and Benullo.

Hank's last words on the phone came back to him.

It was the wife . . .

Benullo's wife, who had died at Briarwood? What did that mean?

Will quickly searched Hank's body again, but whatever evidence Hank might have found on Benullo and Salk had been taken by his—

Killer.

Will flinched again at the implication. But as he was about to turn away, he saw a scrap of paper under Hank's stiff thigh. Will reached for the white, jagged corner and drew a black-and-white photograph from under Hank's body.

Will's knees weakened in shock, seeing it.

It was a photo of the house, before it was burned. The same sagging porch, a marginally better paint job. A young girl sat on the steps, perhaps thirteen, colt-legged in her Catholic school uniform, dark hair and pale skin and deep, haunted eyes . . .

Joanna.

CHAPTER FORTY

Will stood paralyzed in the burned room with its gaping holes in the skeletal walls.

Joanna had come home.

For a moment he felt the presence of her child self in his marrow. This had been Joanna's prison—her private hell.

He felt blistering rage at the knowledge of what had gone on here, the terror and pain and degradation the child she had been had endured. The evil of the house seemed to linger, seeped into the rotting and burned boards, as if the violation of years had scorched the wood.

Joanna coming here. Hank lying dead in her childhood bedroom. *What in God's name was going on?*

The sirens were louder, now blaring outside, then cutting off abruptly in front of the house. Red lights flashed against the broken and burned walls, splashed across Will's face. Will slipped the photo into his coat, forced himself to walk outside to meet the police.

The questions came from far away, as if he were in a dream.

"What were you doing at the house?"

"He called me, asked me to meet him."

"What for?"

"I assumed it had to do with the campaign. He was doing some work for my campaign manager."

"Why there?"

"I have no idea."

"Had you ever been to the house before?"

"No."

Will sat at a battered table in a back office of the Dorchester police station, where the two detectives had suggested they meet to spare him the trip to police headquarters in Schroeder Plaza. He knew them, they knew him. Will's grief and anger were real, and Jerry's story was working; he could see it in the detectives' faces.

Inside himself, Will could barely keep his focus on their questions. He could feel the photo of Joanna against his heart, could hear Jerry's tense query in his mind:

Is Joanna there?

And Hank's last words:

It was the wife . . .

Will was jolted back to the present by the detective's voice. "Did you know he was using?" the older detective, Baronski, repeated. Will took a moment to consider, turning his coffee mug in his hand, though he had no illusions about Hank's death being an accident. He exhaled, spoke heavily.

"I thought he'd kicked it. I know he had for a while."

"Do you know anyone who would want to kill him?"

Will looked across the table at the detectives. This was the moment, the precipice. It was murder now, the time to talk about Salk, about Benullo . . .

The wife . . .

He did not know if it was the thought of the campaign, or Jerry's admonition, or some darker impulse of his own, but without letting his face change, Will spread his hands wearily. "We've put a lot of people away."

The detectives considered him for a beat that lasted a hundred years, then Baronski nodded, stood.

"We'll keep you in the loop."

"I appreciate that."

Handshakes all around.

Will walked through the police station, nodding to uniformed officers as he passed. All he could see was the derelict house, the room with the sagging bed. How could he not have put it together?

Joanna had escaped that place at fifteen, run away from home and lied about her age to wait tables, fending off customers and the night manager. She had never returned, had begun her long, slow crawl up from the gutter—and Joanna's father had died, drunk, in a pool of his own vomit a few years later.

Will fumbled the photo from his pocket, stared down at it, his eyes burning.

What was she doing there?

There was no earthly reason for her to go back. She had never shown the slightest inclination toward facing her past. The opposite: It had been the one subject Will had never been able to touch, for fear of losing her entirely. She'd preferred to eradicate all memory of it from her life.

Will forced himself to open his eyes, to keep walking in the dingy hall.

He knew now that far more than drugs were involved. But what could have persuaded Joanna to go back to *that* house, in the dead of night, was beyond his comprehension.

And whatever the secret was, Hank had been killed for it.

Will felt he was in the grasp of something monumental, a conspiracy far beyond what he had imagined. He almost turned around right there, to go back to tell the detectives everything he knew.

But Joanna.

Is Joanna there?

Had she been?

The thought held him back. He had to protect her at all costs.

But from what?

He walked on blankly. Straight ahead of him in the hall, a uniformed cop stooped at a coffee machine, holding a cup in the slot while it filled. The light above the officer was stark, casting long shadows. Will stopped in the hall, staring. The fiery red hair on the man was unmistakable. The cop from the hospital.

Will stood for a moment with the disorientation of seeing him out of context. The feeling of uncanny significance was strong; he had an impulse to turn and walk away. *Not walk—run, get out, never look back.* . . .

He clenched his fists, moved forward, and spoke. "Officer O'Keefe."

The cop turned, steaming cup in hand. His face was haggard and bleary; he'd lost an alarming amount of weight since Will had seen him at Briarwood.

"I'm Will Sullivan. District Attorney Sullivan."

Will watched O'Keefe's drawn face as he finally made the connection.

"Oh. Oh yeah."

Will extended a hand. The cop looked around as if unsure what to do with his coffee, transferred the cup to his left hand, and shook Will's. His grip was shaky; he was so distracted that he didn't even meet Will's gaze, but Will could see the officer's eyes were rimmed with red from crying.

"I was at Briarwood the night they brought you in," Will offered, increasingly uneasy. O'Keefe seemed to be having trouble following. Will continued carefully, but compellingly. "I've seen a lot of gunshot wounds. To be honest, I thought you were a goner."

A ripple of emotions passed over the cop's features. "We had a baby on the way. Had to hang on, for the kid." Will's skin was now crawling with apprehension as he noted what he knew was not a simple past tense.

"You had to?" he said involuntarily.

The cop's face dissolved. "We lost him."

It was not pity, but some nameless horror that washed over Will like a wave.

The cop rubbed at his wet eyes, straightened. "Sorry." He turned vaguely to go.

Will stood as if paralyzed, then called after him, too loudly: "Did you ever meet a man named Salk?" O'Keefe looked back at him. There was a tremor in Will's voice as he continued, "At the hospital. Were you ever visited by a counselor named Salk?"

The cop looked at him blankly, no recognition on his face. "There were a lot of people . . . doctors and all . . ."

"Your wife never mentioned him? *Salk?*"

"No. I don't know." The cop's face trembled with grief. Suddenly Will could barely look at him.

"I'm sorry," he managed. "I'm truly sorry for your loss."

The officer nodded, staring into his coffee, consumed in private misery. Will backed up a step, turned to leave him. O'Keefe spoke behind him.

"It's not right."

Will turned and looked back at him. The cop stood silhouetted in the glare of the lights above him, a faceless shadow against the light. His voice was hollow, lifeless, sending a chill through Will's marrow.

"He wasn't supposed to die. It should have been me."

———

When Will walked out of the back of the elevator, the dead halls were as gleaming and inexplicably menacing as he remembered in vague, taunting nightmares—and, as always, eerily empty. His brisk pace turned into a jog, then he was running again, racing something unknown and terrifying, his breath hot and harsh in his lungs.

He came to an intersection of halls, slowed to look around him at the three possibilities.

His heart leaped into his throat as he spotted a wall of black: three robed figures standing at the very end of one hall, watching him. Will froze, staring. The center nun held a swaddled baby.

The nuns stepped back in unison, disappearing around the corner with the infant.

Will pounded down the hall, rounded the corner—and halted. The nuns had vanished.

He was looking into a cubistic lobby he hadn't seen before, with red chairs and couches, the color startlingly obscene against the blinding white floor and walls. The red chairs were filled with pregnant women.

He turned away from the ward and jolted. O'Keefe's strawberry-haired wife stood in the hall, hollow-eyed and slim, the swell of her pregnancy gone. Will stared across at her, his breath rasping in his throat.

"Where's my wife?" he said hoarsely. She turned her head away from him.

Will took a step back from her and then stumbled in shock.

Gary stood on the other side of him, just a skeleton with skin, his eyes sunken and burning with dementia.

"Where's my wife?" Will gasped again.

Gary put a bony finger to his lips.

Will turned from them and ran.

He burst through the double doors onto the mezzanine of the skylit marble rotunda. He stopped, staring around him. It was all there: the massive dome; the circles upon circles of balconies; the lush, cascading vines; the giant sculptural hand pouring water down three stories to the fountain below.

No hallucination.

Or a hallucination that had returned.

He strode to the railing and looked down.

The orderly was there below, laboriously pushing his cart up the ramp. Far below, the patient in silk pajamas sat on the edge of the fountain, reaching down for the water.

Will reeled with the impossible sameness of it all. He gripped the iron rail, lunged forward, shouting over space, *"Who are you?"* His voice echoed around him, off the marble walls, through the water splashing from the giant hand. The orderly and the patient never looked up.

Will woke in a sweat, to pouring rain outside the window.

Joanna was beside him, curled into herself in sleep.

Will lay still, staring up at a cobweb in a corner of the ceiling, so faint that it could have been his imagination. His pulse was racing; he felt as panicked as he ever had during the long bleak winter of the hospital, during the very worst of Sydney's illness, haunted by a press of tormenting images. *Joanna in that filthy room. The photo. Hank's corpse. And the O'Keefes' baby . . . the baby . . .*

His thoughts were a vortex of fear and confusion. Sydney was home, they were home, but somehow they had never left the hospital.

He dressed quietly, watching Joanna in the mirror. She was sleeping heavily, not stirring, and there were dark circles under her eyes, like bruises in her pale skin. He moved to the bed and sat beside her on the edge of the mattress, traced his fingers down her bare arms, one, then the other, as he often did to bring her out of sleep. But never before had he been looking for track marks as he did so.

There were, of course, none.

He thought she was awake long before she opened her eyes, as if she were gathering herself, preparing.

She opened her eyes slowly. He smoothed a fall of fine dark hair back from her face, put his hand to her cheek, and their eyes met.

She instantly turned her cheek into his hand, an embrace that hid her eyes from him. He sat with his hand on her hair, knowing the new terror for what it was.

Had she been to the house before him and met Hank there? *Whatever she was hiding, would she kill to keep the secret?*

He swallowed down tears.

Ask her, he thought. *End this. Just ask.*

And then Sydney called from her own bedroom: "Mommy!" Joanna looked away from Will to the clock on her bedstand. "Oh God, Sydney has a playdate at nine. . . ."

She started to get up—he caught her hand. "Joanna." She held his hand, looked down into his palm for a long moment.

The sound of cheery music suddenly blasted from the hall. "No TV!" Joanna called automatically. And slowly, she withdrew her hand from Will's.

"What are you up to today?" she asked with false lightness.

He stood and went to the mirror to do his tie, like an actor feigning stage business. "Meeting Hank," he said, watching her in the mirror. There was not a trace of ambiguity in her face, nothing but simple curiosity in her question.

"Hank? Why?"

It seemed so real.

But you didn't know. She was drugging you, and you never knew.

He concentrated on his tie. "Jerry's using him for some campaign work."

"Mommy!" Sydney's voice was more demanding.

Joanna smiled wryly. "Her Highness." She rose from the bed, stepped to Will, and deftly turned the tie into a perfect Windsor knot. His heart was breaking as she leaned up and kissed him.

CHAPTER FORTY-ONE

The drive, the endless drive. The road, the bridge, the harbor as dismal as the sky. Will called the campaign office from his earpiece and waited, listening to the sounds of Jerry clearing his office, shutting the door.

"The cops were by this morning. Routine questions, so far. Jesus, Will, what the hell?"

"I don't know," Will said flatly.

"Have you talked to Joanna—"

"She's got nothing to do with this." Will's voice was ice.

"Of course not," Jerry assured him.

There was a long silence on the phone. Will stared out the windshield at the curving road. "Cancel everything I've got this week. We'll go—somewhere. I'll let you know."

"Good." There was unmistakable relief in Jerry's voice.

Will looked at his own eyes in the rearview mirror. "And put someone on the house for today. Just until we know more."

"You got it."

Will disconnected and drove on toward Briarwood.

———

The elevator doors closed, surrounding him with his own reflection. The sudden hitch and ascent were far, far too close to the feeling of his dream. Odd enough that he had dreamed of the hospital rather than the burnt house, after finding Hank lying dead in Joanna's old bedroom . . .

He closed his eyes against the thought.

But the hospital was the nexus, and with any luck at least one witness was still here.

The back doors opened behind him. Will turned and stepped through the mirrors, out into the windowless brick service corridors with their dim, soporific light.

The blue line was on the floor, waiting for him. He followed it with the unshakable sensation of being led.

He rounded a corner and found himself in the hall with the soft gray light, face-to-face with the stone statue of the angel on a pedestal, holding up a set of scales.

At the other end of the hall, the grandfather clock ticked.

Will looked at the open door beside the statue—Gary's room—and stepped quietly through.

The room was in shadow from the darkening rain clouds outside the window. A ghost of a man lay in the bed. If anything, he was more emaciated than when Will had last seen him, a skeleton barely covered in flesh. Will spoke softly. "Gary . . ."

Then he froze.

The gaunt man in the bed was not Gary, but Mark—his sensitive, dark-haired lover. But while before Mark had been perfectly healthy, he was now as wasted as Gary had been, cheeks sunken, skin gray and pasty, breathing shallow and labored.

Will felt the room spinning, his entire sense of reality crumbling. "Mark?" he whispered.

Mark turned his head weakly on the pillow, looked up with feverish eyes.

Will moved to the bed, unable to believe what he was seeing. He reached to take Mark's hand . . . and saw the silver AIDS bracelet he had noticed before on Mark's arm. Mark's sticklike fingers barely moved to squeeze Will's hand. Will was overwhelmed with pity . . . and dread. "*What happened?* When did you get sick?"

Mark tried to lift his head. Will bent down to him. Mark's lips worked; Will could barely hear his tortured words.

"Tell him . . . I won't do it."

"Tell who?" Will whispered, though he knew.

Mark looked up into Will's eyes . . . drew a rasping breath. "Tell Gary . . . I told him no."

Mark's head dropped back; his eyes closed. For a heart-stopping moment, Will was sure he had died.

Then Mark's sunken chest moved faintly, in tortured sleep.

Will stood paralyzed, with Mark's hand in his. The radiator hissed as it started up; the sky opened and rain thundered down outside the windows.

"Where is he?" Will whispered. "Mark. *Where*—"

A sharp voice spoke behind him. "What are you doing?" Will turned to see a nurse in the doorway, thin, severe, and efficient. He didn't know her, but he could see her face change as she recognized him.

"Mr. Sullivan? I'm sorry, this patient isn't allowed visitors."

Will looked at Mark and felt as if he were suffocating. "He's a friend."

The nurse bit her lip, shook her head. "I'm sorry. I have to ask you to leave."

Will squeezed Mark's hand and put it gently back on the bed.

As Will stepped out of the room, he turned to the nurse. "What happened to the patient who was in this room before? Gary?"

"He was discharged."

Will stared at her. "Discharged? To another hospital?" He could see his agitation reflected in the nurse's wary face. His voice rose. "Is he worse? Did he die? *What?*"

The nurse actually took a step backward. "I'm sorry—I'm not allowed to give out that information."

She turned and walked quickly away from Will, down the hall. Will stood still.

After a long moment, he turned . . . and found himself facing the angel with the balance. Rain beat against the windows behind him. The grandfather clock ticked.

CHAPTER FORTY-TWO

Will's face in the rearview mirror was as gray as the sky outside the windshield as he drove on a winding street in an exclusive Brookline neighborhood of multimillion-dollar homes. He was unable to turn off his mind, the image of Mark's wasted body in Gary's hospital bed. His rational brain fought for control, spinning out explanations. Mark lived with Gary, they were intimate. Of course it was possible that he'd contracted the disease himself.

A deeper instinct shouted back inside Will's head. *So quickly? To literally trade places with him? To be there in his bed?*

And the alternative is what?

Mark's fading voice whispered: *Tell Gary . . . I told him no.*

Will's mind shied away from the implication. *What you're thinking is not possible. Keep moving. Get the facts.*

He forced himself back to the task at hand, checked the address he'd gotten from Cass against the house numbers moving past.

Sarah Tennyson's house was an Italianate mansion, set far back from the road, behind high brick walls with an automatic security gate.

The housekeeper buzzed Will through the gate when he gave his name. She opened the door to him but stood like a sentry, austere in a formal uniform and tight bun, planted firmly in front of the door.

"Ms. Tennyson hasn't been well," the older woman said apologetically, but she didn't budge.

Will thought instantly of Mark lying in Gary's hospital bed. "Nothing serious, I hope."

The housekeeper gave him an odd look but said nothing. Will pressed.

"It's very important that I speak to her."

The housekeeper finally opened the door and left Will in the immense marble entrance of the elegant home. Chandeliers and Italian paintings graced the hall; a sweeping staircase led upward. Will moved across a luxurious Persian carpet to look over a gallery of opera posters, photos of a glowing Sarah Tennyson onstage in the footlights, draped in extravagant costumes, her lovely face animated with passion.

He turned from the posters as he heard a step behind him. The housekeeper was in the arched doorway, standing ramrod stiff.

"Ms. Tennyson will see you." She paused. "But I must warn you first—she doesn't speak."

Will looked at the older woman. There was something strange in her tone.

"Laryngitis?" he asked, though he knew already it was more.

The housekeeper's voice was flat. "The doctors don't know. She can't speak. She can't sing. Not at all."

Sarah Tennyson met Will in an elegant drawing room dominated by a huge grand piano. She was dressed in peach silk with gold jewelry, her blond hair swept into a French twist: the essence of serenity and grace. She pressed his hand warmly and indicated the damask sofas by an arch of windows, her motions and movements so fluid and expressive that one might not even notice she wasn't speaking.

The housekeeper appeared with a silver tray of cakes and coffee, which Sarah poured herself without a word, her eyes inquiring of Will how she could possibly help him. The housekeeper withdrew, leaving a linen stationery pad and gold pen on the table beside Sarah.

"I'm so sorry to hear of your . . . condition," Will started.

Sarah lifted her shoulders gracefully and wrote on the pad she held in her lap.

There are worse things.

She held up the page. Will read, looked at the singer. She wrote again.

Anthony is well. And Sydney, too?

Will looked at her. "Yes. She's—very well."

Sarah shrugged eloquently. She bent to write.

Then we're both very lucky.

The room seemed colder. Will straightened, looked up from the pad. "Ms. Tennyson, I'm here because I'm trying to find someone I met at the hospital. You spoke with him. Salk."

Sarah looked at him. She shrugged, spread her ringed hands—the picture of puzzlement.

Will tried to relax his jaw, to keep his voice neutral. "Salk. Tall, dark, elegantly dressed. Well-spoken."

Sarah bent to the pad and wrote. The scratching of the pen seemed very loud in the silence of the room. She held up the page.

I don't know who you mean.

Will stared at her. "I saw you talking with him."
Sarah shook her head, wrote:

No.

Will started to speak—she wrote more firmly.

I don't know who you're talking about.

Will looked up from the page, into her face. His voice was very low. "I saw you with him. It was the day before you took Anthony home."
Sarah wrote quickly, in silence. When she was done, Will looked at the note.

You're mistaken. <u>Take care</u> of your little girl.

Will's pulse spiked at the deliberate underlining of the words. Sarah met his eyes, held them.
Outside in the hall, a door slammed, and Will caught a glimpse of a little boy running past the double doors of the parlor.
Sarah automatically rose to call out to him. Her voice came out in a grotesque gargle, a sound Will had never heard or dreamed of—horrible, inhuman.
He sat frozen, unable to believe his ears. Sarah's hand flew to her mouth . . . her eyes were mortified, and frightened.
She looked away from Will's gaze and hurried from the parlor.

Will rose, then stood alone in the magnificent room for a long time, not moving until the housekeeper appeared to show him out.

They walked the gleaming hall in formal silence, past the stage portraits of Sarah, captured in song. At the door, Will turned to her abruptly.

"How long has her voice been gone?"

The housekeeper met his eyes. Her own look was veiled.

"Since the day Anthony came home."

Will walked into the entry hall of his house and closed the door behind him. The grandfather clock in the corner ticked.

Instead of announcing himself, he walked through the double doors into the living room, lit only by the moonlight through the picture window.

Without turning on the light, he crossed to a wall where the sound system was mounted between curved racks of CD shelves. He looked over the CDs, the classical section. He found what he was looking for almost immediately: a collection of arias. He looked down at the photo of a radiant Sarah Tennyson on the case. He and Joanna had bought it after a midsummer night's performance at Tanglewood, the first time they'd heard Sarah sing.

Will put on the CD, and an aria filled the room—from *La Bohème,* Sarah's ethereal soprano, hauntingly beautiful. A gift of a voice, more precious than gold.

A singer loses her voice. Her prized possession. Her son gets well.

His mind reluctantly pressed on with the train of thought.

A cop's wife loses their baby. Her husband gets well.

He heard a step behind him, felt movement, and turned quickly.

Joanna stood, a shadow in the doorway.

She crossed the room and slipped her arms around him, resting her head on his shoulder.

"Haven't heard that in a long time."

Her voice was clear, unslurred, her touch strong, not slack-muscled. No sign of drugs. None at all.

They listened to the music together in silence, the notes shivering through their skins. Finally Will spoke. "You know her son, Anthony, was on the ward." He felt Joanna nod against his coat. "He went into remission a few days before Sydney."

Joanna didn't answer for a fraction of a second too long. "I didn't know. That's wonderful."

Will closed his eyes. "She's lost her voice. It's gone."

Joanna tilted her head to look at him in the dim light from the hall. "What?"

"It's gone. Completely. She can't sing. She can't speak. It's . . . horrible."

The music, Sarah's voice, filled the room around them. Joanna didn't speak for a long time. He thought perhaps she wasn't going to.

"But her baby's well."

Will turned to her slowly, unable to completely believe he'd heard her. "What?"

She didn't answer. Will stared at her. Her face was a pale oval in the dark.

"What do you mean? What does that have to do with it?"

Joanna's voice was steady. "I meant—if her baby's well, I'm sure nothing else matters to her."

Will's blood was pounding in his head. Joanna looked away from Will's eyes, shrugged. Her voice was strangely detached. "It probably doesn't matter to her at all."

The aria crescendoed. Will's thoughts were racing, escalating even as the music sobbed and pleaded.

A singer loses her voice—her child becomes well.
A healthy man takes his dying lover's place.
A woman loses her baby and her husband lives.
A wife dies and her husband becomes mayor.

If one miracle has ever happened in the world . . .

But *miracle* was not the word he was thinking of.

CHAPTER FORTY-THREE

The night was alive, raging with elemental violence. Wind lashed through the trees; the forest pressed in on the dark house, bare branches reaching and shuddering in the whorls of air. Sydney's rope swing tossed in the wind, creaking on its hinges.

A full moon gleamed behind the moving shadows of branches . . . the wide lawn was awash with the pale light.

In the master bedroom, Will jerked up out of sleep. Moonlight spilled across the four-poster bed; the rest of the room was filled with shadows. Outside, a branch scraped against the window in the wind—the sound of the creaking swing from Will's dream.

As his rapid breathing quieted, he looked instinctively to the other side of the bed.

Joanna was gone.

Panic shot through him. Somehow she'd eluded him again.

That vile house . . . that stench of pain and death . . .

Will grabbed for his robe on the bedpost. His eyes fell on the window, caught a flicker of movement.

Below in the yard, the tree swing rocked, creaking . . .

Will tossed the robe on the bed, found his pants, pulled them on.

He hurried barefoot through the bedroom door into the up-stairs hall. The whole house was dark and so quiet that he could hear the ticking of the grandfather clock downstairs.

He paused by the open door of Sydney's room, glanced in to check on her.

The bed was empty.

Will took a startled step into the room, toward the bed—and froze. An ugly, dwarfish figure stared balefully from the corner of the room. Will's heart leapt into his throat . . . then he recog-nized Sydney's Tenniel rabbit, a papier-mâché monstrosity as tall as Sydney.

He breathed out, then his heart vaulted again at a creaking sound from behind him. He whirled—

Sydney sat in her rocking chair by the window, rocking. *How had he missed seeing her?* She looked up at Will. He took an un-steady breath.

"Sweetheart? Where's Mommy?"

Sydney looked out the window. Will moved slowly forward, looking down.

In the dark outside, Joanna stood in the center of the lawn, holding something cradled in her arms. She gazed up at the moon, dark hair tumbled down her back, like a priestess from an-cient days. She raised her arms high—

In her hands she held the little white rabbit.

Will woke sharply—his face taut, beaded with sweat. He looked immediately over at Joanna's side of the bed. She was there, sleep-ing soundly. The room was flooded with a thin, clear light.

———

Will stepped out of the house onto the side porch, his face still lined with tension. The day was cool and breezy, the wind chasing high fluffy clouds across a crystal sky. Will pulled his jacket closer around him. He looked across the lawn . . . his eyes found the tree swing, swaying slightly in the breeze.

He sat heavily on the porch glider. His eyes were gritty, as if he hadn't slept for days, as if he were drugged. His mind raced over the same nightmarish track.

It had to fit together somehow. Sarah Tennyson, Hank, the cop and his wife, Gary and Mark, Benullo and his dead wife . . . and Joanna.

The names and faces were jumbled in his head like some gigantic, terrifying puzzle that had no connection, no sense . . . but Salk . . . and the hospital.

He looked out at the lawn, saw Joanna as in his dream, standing in the moonlight with the rabbit raised in her hands.

Your wife does not care about logic, or reason, or reality. If someone told her she could save Sydney by going out under the full moon and sacrificing an animal, by spilling her own blood, she would do it—without question.

"No," Will said aloud, not realizing that he'd spoken.

"No," he whispered again.

He rose from the swing and walked down the steps of the porch, across the lawn toward the sawhorses that held the rabbit cage.

His breath came shorter, his steps faster, as he neared the rabbit hutch.

He stopped still on the grass and looked through the wire mesh.

The cage was empty . . . the little rabbit was gone.

In the play area beside Joanna's garden, Sydney dug in her sandbox with a plastic shovel, occasionally blowing wisps of hair off her face with a poof of air. She sang softly, breathy nonsense words that she repeated again and again in the same tune:

"Munchee munchee munchee chee . . . munchee munchee mun chee chee . . ."

Will stepped over one of the logs that lined the sandbox. Sydney looked up. Her face was inscrutable.

Will sat on the log to face her, like an awkward giant. He spoke gently, as if the words were fragile.

"Honey, I'm sorry . . . but I think White Rabbit is gone."

Sydney looked at him with her clear gray gaze. "I know." Will stared at her, not trusting his own ears. "He had to go," she explained. Irrefutable logic. She bent her head, back to digging, yellow plastic shovel crunching in the sand.

"He had to?" Will repeated, feeling caught in an unfathomable dream.

Sydney said nothing. She stabbed the sand again, digging deeper.

"When?" he said, more urgently.

She shrugged. "When he had to go."

Will looked down at her blond head, the hair so luxuriantly restored. He closed his eyes for a moment, then said it:

"Sydney, do you still see the man?"

Sydney glanced down, swirling sand with her shovel. Will's throat was so tight, he could barely speak.

"Have you seen him with Mommy?"

Sydney looked up at him reprovingly. "*Daddy*. We're not supposed to talk about the man."

She returned to digging, singing her song under her breath.

So beautiful as they picked flowers in the garden, arranging an elaborate bouquet. Mother and child. A study in contrasts: Joanna so pale and dark; Sydney the alabaster child, silky blond with her gray eyes. No physical resemblance between them apart from their powerful, indisputable femininity.

They left the neat and orderly landscaping of the lawn, moving through the border of trees at the edge of the yard, and walked in the woods along a narrow path, Sydney carrying the bouquet, running ahead after a butterfly, laughing, then running back to Joanna. Birds twittered in the branches above. Beams of sunshine broke through the trees as the woods gave way to a meadow.

Joanna and Sydney emerged from the wooded area and crossed into the rippling meadow, wading into the middle. Dragonflies darted above the grass, sun shimmering on their wings. A stream sparkled beyond.

Will watched from behind the trees as Sydney knelt to put the bouquet on a tiny grave, marked by a flat headstone.

A grave.

He nearly jumped out of his skin as his cell phone vibrated against his leg. He grabbed for the phone and ducked behind a tree to answer it.

"Sullivan," he said, low.

Silence, then a male voice spoke on the other end. "I heard you were looking for me."

Will stiffened . . . but the voice was young, brusque.

"Connor," Will said. He felt an overwhelming sense of relief he couldn't identify. It had to do with science, with order and reason. He leaned to glance around the tree trunk. Joanna and Sydney were walking hand in hand back through the meadow toward the house.

Will leaned heavily against the tree, spoke into the phone. "Where can we meet?"

CHAPTER FORTY-FOUR

It was an Irish pub, one of the dozens within a two-mile radius of the hospital. Dark wood, Guinness on tap, Chieftains on the jukebox, JFK and Bobby Sands on the wall.

Will found Connor in a booth in the back. Without his lab coat, in a fisherman's sweater, a pint in front of him, the young resident seemed more like a college kid than a doctor—except for his eyes. Old.

Will sat across from him, and could see how bad he must look from the jolt of shock in the younger man's eyes. Connor looked away, took a swallow of stout. "How's Sydney?"

"She's great," Will answered automatically, as he had been answering to so many queries for so long. Because on the surface, she was. More than great: Every MRI since she'd left the hospital had shown a phenomenal reduction. Will recited the results of that week's checkup: "Last MRI showed the tumor is barely ten percent of its size before the smart bomb protocol."

Connor was looking at him, waiting for him to finish. Will paused. "There is no tumor," Connor said quietly.

Will stared at him, cold fingers of dread

And more than dread

creeping up his spine. Connor reached to the seat beside him, flipped a manila envelope across the table. Will opened it with his mouth as dry as bone, to find the dark film of two X-rays.

Connor took the films and held them up one at a time against the neon light of a HARP ad on the wall.

"April fourteenth, one oh-seven P.M."

The first X-ray was the familiar image of Sydney's tumor in its most advanced stage. Will felt the old horror, seeing it again: grapefruit-size, massive, malignant.

Connor put down that X-ray and held up the second one. "April fifteenth, two-ten A.M."

The X-ray was completely clear. Organs in their proper places. No sign of the tumor.

In the silence, Will was dimly aware of the music on the jukebox: Celtic violins. Inside, he was reeling. But not with shock. It was confirmation. It was relief. It was madness, but it was what he had somehow known.

"It disappeared—that night," Will said hoarsely.

Connor frowned at him. "No. Of course not." He paused, and Will sensed he was struggling with himself. Then Connor continued, his voice flat. "The chemo had been working from the start. It's a lucky thing we went in that night because we could have killed her with the endostatin."

Will could barely breathe. "I don't understand."

"The tumor was shrinking all along, and we didn't know until that night. Someone fucked up the files." Connor looked away from Will. "It was me."

Even though he did not immediately process what the young doctor was trying to tell him, Will's response was automatic, heartfelt. "That's impossible."

"There was another Sullivan. Also with a pancreatic blastoma. Two different people—two different hospitals—but the labs are processed through the same radiology department. They're all digital files, and I was picking up the wrong Sullivan."

"Over a period of weeks?" The prosecutor in Will laughed in disbelief. "How did you do that?"

Connor looked haunted. "I hadn't been sleeping. I'd been doing NoDoz every day to stay awake. I might as well have been doing meth. I got the wrong file once because I didn't double-check the names, and after that I was putting in the wrong lab numbers. "This is a picture of a different patient." He tapped the first film.

Will was more and more incredulous. "*I've seen* the X-rays—they've been showing us X-rays of the tumor shrinking—"

"That's what I'm saying. They found Sydney's films. That's what they've been showing you these past two months—the X-rays and MRIs of the tumor receding. But it all happened before. They changed the dates on the files, that's all."

"*What*? Why?" Will was struggling to fathom it.

"They were scared shitless that you'd sue the hell out of them if you found out that we were aggressively treating your daughter for a tumor that was already in rapid remission."

Will reeled with thoughts on all levels. *Joanna bathed Sydney. She knew every inch of her body, watched every change in her like a hawk. She must have known the tumor was gone. All those days that she was taking Sydney to the hospital for her checkups—she must have known it was a charade.* That part he could believe. It was part of what he had sensed—and denied. But there was so much more wrong with the story.

"I don't believe it," Will said. "There's no way you could make a mistake like that. Once, maybe. Not weeks at a time—"

The young resident shook his head. "I was there when Mankau opened her up." He laughed humorlessly. "Oh man—I've never seen him so freaked. They were calling out all the technicians.

They wanted genetic tests—they were even thinking a twin . . . and this is not just any kid, mind you, but *Will Sullivan's* daughter. And then they found the missing files."

"I don't believe it," Will repeated.

Connor took a deep swallow of Guinness. "I fucked up. They let me off easy—gave me a transfer with a recommendation if I didn't talk. They'd have done anything to avoid a malpractice suit. Christ, the DA's kid. . . ."

Will spoke sharply. "Then why are you talking to me now?"

Connor looked away. "When I heard you were looking for me, I figured I owed you the truth." He drank again.

Will's thoughts were leaping . . . from Mark in Gary's hospital bed, to Sarah's grotesque voice, to Joanna sitting in the dark in the burned room on the filthy bed. On the jukebox, the Irish violins played . . . an unearthly sound. His whole world was crumbling. But it was time to stop pretending. He knew now. He closed his eyes, then opened them and leaned toward Connor.

"They're not letting you off easy. They faked those films they 'found.' They set you up."

"Set me up?" Connor looked bewildered.

Will looked up at the neon HARP sign on the wall. "They fed you a line, Connor." Will looked down at the two films under his fingertips. "These are the right films. The tumor did disappear. The night of the operation."

The young doctor stared at him. "What—you mean, Mankau took it out? No way. I was there—"

"I mean, it *disappeared,*" Will repeated. The two old guys at the end of the bar looked their way. Will lowered his voice, leaned across the table, speaking quickly.

"That's what they're covering up, and they're using you to do it. You know you didn't fuck up the numbers. I don't care what you were on. It wasn't you. There's something else going on here." Connor was watching him warily, but with a visible flicker of hope. Will plunged ahead. "What do you know about the Tennyson case?"

"Anthony Tennyson?" Connor looked perplexed, but recited the diagnosis. "Advanced lymphocytic leukemia. I wasn't on his team, though—"

Will cut through him. "He went into remission, too. Just before Sydney did."

Connor's reaction was automatic. "Lymphocytic leukemia has an eighty percent remission rate in children who receive proper treatment." The same statistic Cass had quoted. But Will knew statistics had nothing to do with it. He looked up again at the neon light of the HARP ad on the wall. The music seemed alive. He closed his eyes briefly, then went on.

"Anthony's mother has lost her voice. She's been completely speechless since she brought him home from the hospital. We're talking about Sarah Tennyson, the opera singer. She lost her voice, and her son got well."

Connor stared at him, bewildered. "I don't get it. What's the connection—"

Will spoke over him, faster and faster, the intensity of a madman. "There's another patient who was dying of AIDS. He's been discharged. His lover was completely healthy, and now he's dying. And the cop I told you about—O'Keefe. He lived, but they lost the baby—"

"Jesus, what the hell—you're not making sense."

Will stopped. Connor was looking at him with real concern. Will stared down at the films of the tumor: there . . . and then gone. He pulled himself back to the present and leaned in toward Connor.

"The hospital lied to me, and they lied to you." Will saw the struggle on Connor's face and pressed the only advantage he had. "I swear to you, something else happened the night of Sydney's surgery. We need to find out what."

CHAPTER FORTY-FIVE

Somewhere in the bowels of the hospital, Will stood guard in the corridor outside the records room, keeping a wary lookout.

The door opened and Connor slipped out, carrying a stack of patient charts. He closed the door behind him, spoke low. "These are active charts. Anyone could ask for one at any time. We need to make this fast."

Will fell into step with him, and they walked quickly down the gleaming white hall, their nervousness making them as conspicuous as if they were onstage. Will could feel the hospital around them, alert, watching, listening . . .

He forced himself to stare straight ahead, trying not to sweat.

Connor opened the door of a cubbyhole of an office, and they both ducked inside.

Will hovered in the cramped doctor's office, watching through the slightly cracked door as Connor stood at the light board, putting up one X-ray after another, frowning.

Will peered out the gap in the door. The sense of presence was ominous, but the long hall was empty: closed office doors up and down both sides. Then a group of doctors in white coats moved around a corner, coming straight for him. Will stiffened, eased the door almost closed, holding his breath.

The doctors came closer, closer . . . a wall of white . . .

. . . then passed by without noticing the open door. Will exhaled. He closed the door silently, turned back into the small consulting room.

Connor's gaze was fixed on the X-rays on the light board, his fingers moving over the dates at the bottom of the films. Finally he shook his head, his face bleak with disappointment. "There's no discrepancy about these dates. I can't see that these films have been altered in any way. It's just what I told you—Sydney's tumor was shrinking with the chemotherapy. She was in remission long before the night of the surgery."

Will saw Joanna on the lawn, holding the white rabbit up to the moon. He closed his eyes against the thought, spoke sharply. "The others . . ."

Connor glanced over the open medical charts spread out on the desk. Will could see the names: ANTHONY TENNYSON, GARY DOWLING, MARK WEINBERG.

"Nothing. Look. Anthony Tennyson: remission after daunorubicin, tioguanine, and cytarabine therapy for lymphocytic leukemia. I can show you hundreds of case files of remissions engineered by this very treatment."

The young doctor picked up the next chart. "Gary Dowling has been on aggressive treatment with antiretroviral cocktail drugs for the past year—"

Will's voice exploded in the small room. "Jesus Christ, man, you're not listening to me. One day he was *dying,* the next he's been released from the hospital!"

Connor flinched but kept his tone neutral, professional. "He was admitted to the hospital with pneumonia. Looks like it was touch and go there, but it's hardly unusual to recover from pneumonia."

Will felt he was fighting against something ineffable. "You didn't see. His partner was perfectly healthy, and now *he's* the one in Gary's bed."

Connor picked up another chart. "Mark Weinberg's medical record states he's been HIV positive for ten years. The stress of his partner's illness probably pushed his immune system too far, and he developed full-blown AIDS. It happens all the time—"

"And Sarah Tennyson losing her voice?" Will demanded.

Connor shrugged. "Psychosomatic. I've met Ms. Tennyson. She had intense guilt about not having picked up on Anthony's leukemia earlier because she was so involved with concert tours. Maybe she made herself lose her voice so it wouldn't be an issue anymore—"

"O'Keefe's wife lost their baby—"

"Miscarriage from the stress of—"

Will grabbed the charts and hurled them at the wall. Papers and X-ray films hit the plaster, falling and fluttering to the ground. Connor took a startled step back, was stopped by the edge of the desk.

Will ground out hoarsely, "I saw. I saw. A limbless man. No arms. No legs. Just stumps. I saw him with arms and legs. I saw him walk." His hands were shaking. Connor was immobile, watching him like a diagnostician. Will closed his eyes. "These things that are happening . . . they're not—possible. . . ."

When Connor spoke, it was quietly. "What was this man wearing?"

Will turned to stare at him.

"Pants? Long sleeves?"

The buzzing in Will's head was very loud. "What are you saying?"

"I'm saying we do incredible prosthetic work here." The young doctor hesitated, then corrected himself flatly, "I mean, *they* do. You wouldn't believe what's been developed just since the war's been on. So, you saw a guy with clothes on over his new arms and legs."

Will had a flash of the limbless man in the corridor at night, struggling to walk on metal stems. For a moment, his whole sense of reality shifted again, like sand under his feet. Could Connor be right—he'd seen the limbless man with prosthetics and his mind filled in the blanks? Had he bought into a mad explanation for everything he'd seen, some kind of metaphysical shell game? His mind scrabbled frantically for purchase.

Across from him, Connor was shaking his head, staring at the scattered files on the floor. "Every one of these cases has a perfectly sound medical explanation. There are no miracles here. People do get well. I've seen it before. I've seen it all before."

Will's thoughts were veering crazily. Of course there was no trace. Salk had been clever, hiding his tracks, leaving no evidence in the eyes of the outside world. There were no miracles, no strange exchanges, no bartering—no bargains.

Bargains.

Because that's the word you're thinking of, isn't it? Bargain. Deal. Price.

"No. No. *No.*" Will was speaking, then shouting. "You don't know anything. I know what's happening. *I know.*" He could feel the blood in his face, his heart pounding out of control. Connor was backing away from him, holding up his hands, speaking carefully.

"Okay. Okay. Tell me. Can you tell me? Let's talk about it."

Will could see his own madness reflected back at him in the young doctor's eyes. He laughed. He moved back, found the doorknob behind him, and slammed out the door, with Connor calling behind him, "Sullivan. Don't—wait . . ."

Will was already striding into the gleam of the halls, startling a passing young mother and her eight-year-old girl, who carried a bouquet of flowers as they walked toward a room. They both looked at Will warily, and he laughed harshly, unable to stop himself. The mother took her daughter's hand, hastening her away. Knowing that he was on the verge of exploding, Will moved through a pair of double doors to get away from them and found himself on a long glass bridge.

It was the same bridge he had walked with Salk that strange, endless night. The bridge where Salk had asked him something that now danced in the corners of his head. *What are you willing to do?*

But Salk didn't exist. Everyone said so.

Will felt his mind splitting in two: There was a reality he used to know, Connor's reality, a rational, scientific world; and the one in which he was now living, which couldn't exist but did. Or it was madness.

At the junction of the bridge, seven dwarves turned a corner and marched toward Will in lockstep precision.

Not children this time, but grotesque, distorted little men, moving on Will, menacing . . .

Will shut his eyes, shuddered. He forced his eyes open again.

The dwarves marched past him through the double doors at the end of the bridge. The last dwarf turned to look at Will before the doors closed on them.

Will whispered aloud, "You don't exist. Everyone says so."

And then he stopped, remembering.

Mark in his hospital bed—*Gary's* hospital bed—whispering choked words. *Tell him . . . I won't do it. Tell Gary . . . I told him no. Him.*

Will walked out of the back of the mirrored elevator, walked alone down the gray corridor with the softly ticking grandfather clock, past the statue of the angel, toward Mark's room, formerly Gary's.

He halted in the doorway, looked in . . . and his heart dropped.

The room was vacant, the bed empty, stripped, the light above like a spotlight illuminating a painting.

Will sagged . . . and found himself to his vast surprise starting to make the sign of the cross. He laughed shortly, stopping the gesture. After a beat, he spoke softly to the empty bed: "I'm sorry."

He turned away from the doorway, walked slowly in the hall, past the clock with its swinging pendulum, the quiet, inexorable ticking.

He turned the corner.

At the end of the hall, two orderlies were wheeling a gurney toward an elevator. A body lay on the gurney, covered with a sheet.

A tall, blond young man walked beside the stretcher in the slow, dazed pace of a funeral mourner. Will slowed, watching. There was something familiar about the young man's profile, the male-model features.

The orderlies wheeled the gurney into the elevator. The blond young man stepped in after.

Will bolted down the hall toward the elevator, thrust his hand in between the doors just as they started to close. He squeezed himself into the elevator.

On the gurney, between the orderlies, a gaunt man's corpse lay covered with the sheet. A hand hung over the side, stiff, shrunken . . . seeming to reach out in a last desperate gesture. Will looked down—fixed on the silver AIDS bracelet on the wrist.

The blond young man stood by the stretcher, looking down at Mark's body. His eyes were red from crying, but he was gorgeous. Taut muscles, a screen actor's charisma—the picture of youth and vitality.

Gary.

Gary, whole and radiantly healthy, a good eighty pounds

heavier. Will stared at him in disbelief, taking in his transformation, his golden-boy glow.

Gary lifted his gaze from the stretcher—and saw Will. His eyes widened in fear. He spun to the control panel, but the elevator was already descending.

The elevator moved silently down, and down. Will was locked on Gary. The orderlies stared straight ahead. No one spoke.

The elevator stopped; the back doors opened. Will looked through to a long refrigerated room with cadavers on slabs, row upon row of corpses. The morgue.

The orderlies started to wheel the stretcher with Mark's body out of the elevator. Will kept his eyes fixed on Gary.

While the gurney was blocking Will, Gary turned and dashed out of the elevator. Will stumbled around the edge of the stretcher, grazing his thigh, and raced after Gary.

The younger man dodged through rows of slabs of naked bodies. Will dashed after him, leaving the orderlies shouting behind them.

Gary shoved a table in front of Will. Will took the edge in his gut, half fell over a slab. He caught himself on his hands—found himself staring down into the wizened face of a female corpse. He recoiled from the touch of dead gray flesh, the cloying smell of formaldehyde. Death. Death everywhere. Will's flesh was crawling, his mind screaming. He pushed away and spun, looking around him.

Gary was gone. Doors at the end of the room swung shut.

Will crashed through the swinging doors into another long, windowless room: the pathology lab.

Mindful of the orderlies behind them who might follow any moment, Will turned and shot the inner bolt on the doors. He moved into the room between rows of display cases and halted, staring around him at specimen jars and glass tanks, all lit from below to illuminate grotesque specimens of tumors, diseased body parts, heads with harelips, rows of fetuses with encephalitis, pinheads, conjoined twins . . . like a nightmarish carnival sideshow. The stink of formaldehyde was sickening in his nose.

He turned between the cases—and saw Gary, distorted through the glass tanks at the far end of the room.

"Gary!" he shouted.

Gary leapt at the back door and pulled at it, trying to open it.

Will raced through the lab cases and seized the younger man, yanking him away from the door. The two men slammed against a shelf. Jars fell to the floor, exploding on the concrete, cascading glass shards and human viscera.

Will lunged and grabbed Gary, shoved him up against a wall, smearing blood from his cut arms against the white surface. Both men were panting; Will's eyes were crazed.

"Mark's dead. It should have been you."

Gary twisted in Will's grip. He slurred his words slightly; Will could smell gin on his breath. "You're crazy. I don't know what—"

"You were a dead man. Now you're healthy. *How?*" Will's hands dug into Gary's arms. *"Who is Salk?"*

The veins bulged in Gary's neck as he shouted, "Shut up!"

Will was startled into silence. He could see Gary was trembling, on the verge of tears. Will seized the moment, spoke low. "Mark said, 'Tell Gary I won't do it.'"

Gary's face shuddered; his eyes darkened with pain. Will's eyes bored into him.

"He wouldn't—but you did, didn't you?" Will ground out. Gary struggled in Will's hands, to no avail. Will held him against the wall. Gary went limp and sobbed aloud, once, snot running from his nose.

Will's fingers dug into his flesh. "What did you do?"

Gary bolted upright, malice burning in his eyes. "You've got your little girl back, don't you?"

The words hit Will like a sledgehammer. Gary met his gaze

and said softly, "But that Italian kid didn't make it, did he maybe you don't want to know. Maybe you just want to leave this alone—*Daddy*."

The jars with their hideous contents loomed above them. Will was suddenly cold with fear. Gary seized his chance, twisted out of Will's grip, and lurched unsteadily out the back door.

Will made no attempt to follow him. He stood among the glass cases of diseased body parts, unable to move.

That Italian kid didn't make it, did he?

Will stood in the cool green hall, looking up at the wall of glass out into the garden. He had come without realizing where he was going and been standing for he had no idea how long. The garden outside was lush and green, but through the glass wall Will saw it frozen in drifts of snow, a black sky above studded with stars as cold and brilliant as diamonds.

A voice spoke from very far away. "May I help you?"

Will turned.

A chaplain stood in the doorway of the chapel, the same round-cheeked, salt-and-pepper-haired one he had seen before, all that time ago.

Before Salk.

"Would you like to come in?" The chaplain held open the door invitingly. Will looked at the threshold and did not move forward.

"What is it?" the chaplain asked kindly.

Will looked at him from across an abyss, unable to say the words. He turned in silence and walked out the glass doors into the garden.

He walked under gray skies on the gravel path between the trees, through the line of statues, pebbles crunching under his feet like snow. All around him was green; Will could see only white—drifted snow and frozen trees. The marble figures looked down in pity. He stepped around the last pedestal and looked down the path toward the stone bench underneath the overhanging willow tree. The bench was empty.

A breath of wind shivered through the garden as Will approached the bench, eyes fixed on the towering angel across the path.

He spoke only once.

"Salk."

There was no answer.

Will dropped his head and for a moment saw a black velvet glove in the white, white snow. Then his eyes focused. Not a glove, but a glint of metal in the green of the grass. He stooped to pick it up, felt the eerie weight of familiarity in his hand.

A silver rosary.

CHAPTER FORTY-SIX

Fog drifted in the narrow streets of the Italian district, blanketing the sidewalks and tiny yards, deadening sound. The street was deserted; nothing moved, not a car, not a soul, as Will got out of his BMW.

The brick apartment building in front of him seemed to bulge at its seams from the sheer density of tenants. Kids' bikes and laundry racks cluttered the wrought-iron balconies, toys lay scattered in the dingily carpeted halls. The stairs were redolent with cooking smells.

The hefty woman who let Will into the fourth-floor apartment led him down a short, dark hall and showed him into a small living room dominated by a table displaying a grainy blown-up photograph of the dark-haired little boy whose bed Will had seen Teresa praying over. On the lace tablecloth were dozens of lit candles and several vases of fading flowers; more, smaller photographs; toy cars and soldiers: an altar to the Marinaros' lost son.

The stocky Italian father Will had seen in the surgical waiting room was slumped in an easy chair, haggard and listless. A pall of depression hung over the room. The father barely looked up as

Will stepped through the doorway. Will somehow found his voice.

"Mr. Marinaro. I'm Will Sullivan. I'm looking for your daughter, Teresa. She—knew my wife in the hospital."

After a long moment in which Will thought the other man was not going to respond, Mr. Marinaro looked up. His eyes were unfocused.

Will said softly, "I am so sorry for the loss of your boy."

Tears pushed at the burly man's eyes. Gary's voice whispered viciously in Will's mind. *You don't want to know, Daddy.*

Will swallowed, hurried on. "It's very important that I speak with Teresa."

The father slowly turned his head back to look at the photo of his son. He did not speak.

Will glanced at the altar and felt a hole in his chest. Gary's voice whispered again:

You've got your little girl back, don't you?

Will felt movement in the doorway behind him and turned quickly. He drew back as he saw the black garb of a nun. Then his eyes focused and he recognized the tiny dark grandmother from the hospital, dressed all in black, hovering in the shadows of the hall. She looked him over with piercing eyes, then nodded once, beckoning with a clawlike hand.

"Vene."

She took him down the dark, narrow stairs without a word. Outside, she walked in the street through the fog, surprisingly spry with her short steps and ankle-length skirt. Will followed as if in a dream, abandoning himself to her hands.

She stopped at the street corner and pointed. Will could barely make out a stone church through the fog, the cross on the steeple obscured in gray.

He turned to thank the grandmother and was unnerved to find her gone, already disappeared into the mist.

Will moved into the hush of the shadowy church. He stopped on the tiles, eyes straining through the dark. Above him was stained glass, oil paintings in Renaissance style, graphically bloody, on the walls. Pinpoints of light glowed from red votive candleholders in recessed altars. Somewhere in the vast darkness, water dripped in a fountain.

He was hit by a wave of sense memories from his childhood: the rush of cool air, the smell of incense and candle wax, the mystery of the Mass, the cryptic power of the Latin liturgy his father had always preferred, the awe—and fear—Will himself had had of the black-robed priests and nuns in their celibate severity.

He shivered and turned away toward a side altar. In the shadows of the nave, Teresa Marinaro knelt before a statue of the Pietà, Mary holding the dead Christ. Candles flickered above the girl's head as she murmured, an obsessive stream of prayer.

Will approached slowly, his steps hollow echoes on the tile. "Teresa."

Teresa did not look up, did not look at him.

Will knelt awkwardly beside the girl, the gesture uneasily familiar. The candles wavered above them; Teresa's eyes shone with reflected candlelight. She seemed not to know he was there. Will spoke through numb lips.

"I was at the hospital . . . the night your brother died." He reached into his pocket, drew out the silver rosary, and extended it to her. She took it, looked down at the glistening beads, then up at him. Slow tears began to course down her face.

Will spoke with difficulty. "My daughter was sick, too. She's . . . well, now."

Sydney is well, now. . . .

Teresa looked up to the statue and shuddered with silent sobs. Will pressed on. "There was a—man—there that night."

Teresa stiffened. Her face was pale in the dark. "You should not speak of him."

Will felt a knife of dread. "Please. Tell me. It's my wife and little girl—"

"You know," the girl said slowly.

"What—"

"You know what he is."

The candle flames elongated, casting tall shadows on the wall. Will whispered hoarsely, "Please . . ."

Teresa shivered, but after a moment she began to speak, her slender fingers slipping over the silver beads of the rosary.

"For days I see him watch me . . . in the halls . . . he stands in the door when I pray."

Will closed his eyes.

Salk standing, dark and graceful in the door of a hospital room, looking in at the boy in the oxygen tent. The votive candles around the bed flattening as in a rush of wind at his presence.

"The night they operate on Paolo, I go to the church."

Teresa in the garden corridor, about to step through the wooden doors of the hospital chapel . . .

"I see him . . . waiting."

Teresa hesitating at the door, turning . . . to see Salk standing beside the glass doors to the garden, watching her.

Teresa continued to speak, but Will barely heard. He knew. He could see it all.

Salk walking through the doors into the garden. Teresa hesitating . . . before she followed . . . moving slowly on the path, under icy trees . . . stopping beside the statue of the angel. Under the willow tree, Salk standing beside the bench, darker than the darkest shadow, waiting.

In the candlelight of the church, Teresa was trembling, barely able to speak, the rosary clutched in her hand. "He say he will heal Paolo." She sobbed aloud but wiped her face, lifted her shaking chin. "I say—only God can heal. I tell him *no.*"

Will swallowed, against her pain and his own dread. "And your brother died. . . ."

Teresa looked up at the statue of the Virgin. "He is with God now. It is God's will."

Will looked up at the painted alabaster face: cold, lifeless. He forced out the words.

"My wife . . . did you see my wife that night?"

Teresa looked at him.

In the garden, Teresa backs away from Salk, turning to flee and brushing past someone on the path . . .

Joanna.

She stands in the garden, twisting her velvet gloves in her hands, looking toward the willow tree . . .

. . . where Salk now stands alone. His eyes lock with hers . . . in a chilling understanding.

The light of the moon glitters on the snow . . . as the black glove falls . . . and Joanna moves toward Salk's outstretched hand . . .

Will stood in the candlelit church, waves of horror breaking over him. He pulled himself away from his vision, to find Teresa watching him.

She reached suddenly, took his hand, and pressed the rosary into it earnestly. "Only God can heal."

She held Will's eyes, then turned from him . . . and walked into the dark.

He stood alone in the vast emptiness of the church. There was no sound but the whisper of water, the slow drip from the holy water font. After an eternity, he turned—and saw in the transept opposite him the rows of confessional booths. A green light was on over one, indicating it was open for confession.

Will's heart thudded in his chest. He forced himself forward across the tiled floor. He stopped in front of a carved door, opened it, and stepped in.

Inside, he pushed the door shut, closing himself in with a soft click, and knelt in the closet before the screen, smelled the vestiges of perfume, sweat, and despair. His own first confession came flooding back to him: the fear of being shut alone in the tomblike darkness with a God he did not understand, the whole mystery and terror and insanity of the church.

The grate slid open with a clack. "In the name of the Father, and of the Son, and of the Holy Spirit," a deep voice intoned.

Will clasped his hands together, and the words tumbled out of him, automatic, without thought. "Bless me, Father, for I have sinned. It has been"—he groped for a sense of time—"fifteen years since my last confession." The silence seemed to thicken disapprovingly. "I have . . . I fear . . ." Will swallowed.

Joanna in the snowy garden, facing the angel . . .

"Father, someone I love . . ." His voice broke. "I fear for her soul."

Joanna on the lawn, lifting the little white rabbit to the moon . . .

"I believe she has made promises—to—"

Salk on the bench beside Joanna, touching her thigh . . . Salk's eyes, black as coal, burning in the dark . . .

Will's voice twisted with agony now. "That she has sold—herself . . ."

The madness of what he believed surrounded him.

"Help me," Will whispered. "Help her."

After an eternity, the voice came from behind the screen. "My son. Our Savior destroyed the Devil through His death. Confess your sins, and she must confess hers, and Christ our Lord will pardon you. Pray with me now: 'Oh, my God, I am heartily sorry that I have offended Thee, and I detest all my sins, because I fear the loss of heaven and the pains of hell . . . '"

The voice continued in the dark of the booth. The door stood open.

Will was gone.

CHAPTER FORTY-SEVEN

The moon appeared and disappeared through windblown clouds above the forest.

Will stood on the lawn, looking toward the dark house, the one lit window in the upstairs bedroom. He turned and stared up at the moon through the moving fingers of clouds.

He walked off the lawn into the row of trees edging the property at the back of the yard, feet crunching on the leaves as he moved through the woods, deeper and deeper into the dark. A fragment of a story whispered through his head, Joanna's voice: *You cannot find him . . . for he lives in a castle which lies East of the Sun and West of the Moon.*

Out of the trees now, he waded into the long grass of the meadow, following the moon path . . . and stopped in front of the tiny grave with Sydney's bouquet of fading flowers.

He knelt on the earth, pulled up the flat rock that served as a headstone, and used it to dig into the ground, sharp, increasingly desperate jabs at the earth. He hit a hard surface, threw aside the stone, and clawed at the dirt with his hands. His fingernails scraped and tore on something unyielding, and he felt around its corners, discovering a wooden rectangle in the earth. He pulled a toy box out of the ground, cascading streams of dirt, and recoiled slightly.

He could already smell it.

He set down the box, brushed it clean . . .

. . . opened the lid . . . and looked down in the moonlight.

The box was lined with a silk scarf. And nestled in the folds was the putrefying body of White Rabbit. The bunny's small abdomen was distended almost beyond recognition by a grapefruit-size tumor.

Will staggered to his feet, gagging, reeling . . .

Joanna on the lawn, lifting the rabbit to the moon . . .

Will turned in the field, wiping his hands compulsively on his trousers, gasping for breath.

Behind him, a shadow appeared in the moonlight. He sensed a presence and whirled.

Joanna moved out along the path of light spilling across the meadow. Will faced her. His heart was pounding so, he could barely speak.

"What have you done, Joanna?"

She was silent, her face pale as the moon. The wind whispered through the rippling grass.

"Tell me . . . it's not true."

"I had to."

Just like that. Admission. She hadn't even raised her voice. She stood in the moonlight, arms at her sides. Will choked out hoarsely, "My God . . ."

Her words tumbled out in a sudden rush of relief. "I wanted to tell you. But I know you'd never believe . . . you'd never let me . . ." Her voice dropped, intimate, confiding. "It was so clear, Will. He said he could save her. There isn't anything I wouldn't have done." She spoke in wonder, her eyes black and shining in the dark. "And now Sydney's well. The tumor's gone. She'll never relapse. It's total, permanent remission."

There was a strange, feverish excitement in her face. "She's going to live to ninety . . . she'll have two children . . . and grandchildren . . . she'll be a senator, Will. I've seen it. He showed me."

Will fought waves of nausea. It was all completely mad. "What is it? What did you promise him? What are you giving him?"

Joanna's eyes blazed in the darkness. "We're not going to talk about that. Ever." Will took a step back at the force of her fury.

She faltered and looked away. "It . . . doesn't matter. I can do it. It's . . . just like a dream . . ." She trailed off, a little blank. She touched her burned arm without realizing, her face in shadow. "I can do it."

Will could feel his mind cracking. "This isn't real. . . ." He didn't know if he'd spoken aloud. The grave of the rabbit was a black hole at their feet.

Joanna was talking again, a bright, nervous babble. "He'll never go near Sydney. She doesn't even remember—"

Will paced a few steps in disbelief. "You expect—him—to keep promises?"

"He will!" she said savagely. Her face trembled. "It's our deal. We can be together, we can be a family—everything can go on just like before."

"Except you go off to him at night." He could barely control his fury. "If you think I'm going to let you—"

Suddenly it was Joanna in a rage, as he'd never seen her, advancing on him, her face alight with something terrible. "Oh yes, you're going to. You have nothing to say about it. It's Sydney's life. Do you understand? We have no choice. If I can do it, you can.

And you will. Or I'll take her and go now, tonight. You'll never see either of us again."

Will staggered, struck to the core. He knew beyond doubt she meant it.

Then her voice broke, and the fury was gone.

"I made a deal, and I'm keeping it."

A tremor passed across her face. She whispered painfully, "Forget you know. It's just a dream. That's the way it . . . has to be. I promise you . . . it doesn't matter. I love you."

Will stood like a statue as she touched him, kissed him. When she drew back, her eyes were dull. "I have to go now."

Will's face turned hard, deadly. "No. Never. Never again."

"You can't stop me—"

Will reached and grabbed her arm, pulled her into his arms. She fought him, struggling against his body as he pinned her, holding her tight against him, kissing her mouth . . .

. . . and she was sobbing, clinging to him, tears on her face, her mouth opening under his, kissing him desperately . . .

Suddenly her eyes widened and she staggered, looking up at him in confusion.

He held an empty hypodermic needle in his hand.

Joanna's knees buckled . . . and he scooped her up in his arms as she fell.

Will stepped into the house, Joanna's limp body gathered in his arms. The cat watched from its chair as he kicked the door closed behind him and carried her up the stairs.

In their bedroom, he laid her gently on the bed, brushed her hair back from her face, and bent to kiss her.

When he stepped into the door of Sydney's room, she was sleeping peacefully in bed, surrounded by stuffed rabbits, the soft glow of the night-light on her face.

Will stood looking down at her for a long time. Then he turned and walked out the door.

The grandfather clock ticked mechanically as Will walked down the stairs . . . and out into the night.

CHAPTER FORTY-EIGHT

He drove, with trees along the road swaying in the wind, across the bridge, across the bay, the luxurious shell of his car around him, the road in front of him with its shining yellow lines. In his mind, he was in the hospital, in the dim service corridors. *The long brick halls branched out in all directions, elongated, stretching out into infinity; the red line on the floor was bright and malevolent, pulsing like an artery.*

Will drove

and he walked in the hospital. He could hear his own blood pounding in his ears . . . the hospital itself seemed to be throbbing, the walls breathing slightly like the fleshy walls of lungs . . . like the throbbing of an engine, like the steady beat of a heart.

There was no reality anymore. But then he'd known that for some time.

He drove, and the road curved, *and the pulsing red line twisted through corridors, around corners, past the open doors of patient rooms. He knew all too well what was beyond those doors now, in the shadowy cubicles. Visions out of nightmares.*

In the basement morgue, Gary stands between the slabs of cadav-

ers, taking off his clothes. He sits naked on the edge of a marble slab covered with a sheet, then lies back onto the cold stone. The sheets next to him rise, and Mark's rotting corpse leans over to embrace him. Gary's head arches back in a silent scream.

Will drove, and walked.

In a delivery room, the cop's pregnant wife strains on the table . . . shrieking in labor.

The three nuns crowd around the table, attending. The cop's wife cries out as the baby spills out of her, bloody and wailing.

The tallest nun reaches to pick up the crying newborn child . . . and the cop's wife begins to scream in a different kind of agony. The figure in the black cowl is the grotesque nun with the gnarled face and black coal eyes. The monstrous nun folds the baby into her habit, and the three sisters creep from the room . . . as the cop's wife begs and screams.

Will turned his head, and walked on

past another door of another room, where Governor Benullo's wife lies in an impossible puddle, barely recognizable as human, her body turned to soup by her disease . . . the price of her husband's ambition . . .

Will clenched his hands on the wheel and drove . . . keeping his eyes straight ahead. *All around him were the sounds of whispers and shrieks . . . pleas for help . . . the sound of buzz saws and cracking bones.*

Will reached for the turn signal and drove off the ramp.

In his mind, he pushed through the double doors onto the mezzanine of the rotunda . . . with its lush, malodorous vines crawling over the balcony.

The ramp spirals down vertiginously . . . down . . . down . . . the red line pulsing like blood.

Down . . . past the orderly struggling up the path . . . every muscle straining to push his cart.

Down the sickening loops of the spiral . . . down to the garden

where the patient in silk pajamas kneels on the edge of the fountain,
both arms vainly extended toward the water, forever a fraction of an
inch beyond his fingers.

Down into the garden . . .

Where the ramp stops abruptly at a splintered wooden door.

Will put his foot on the brake and stopped the car in front of
the condemned house on Cabarrus Street. He looked up at the
burned house with its splintered wooden door . . .

. . . then opened the car door and stepped into the dark.

CHAPTER FORTY-NINE

Will moved through the doorway at the end of the dank and rotting hall and stood in the small, dim room, staring at the sagging iron-frame bed. Sickly moonlight glowed faintly through the filthy windowpane. The wind whispered along the eaves.

He felt the presence even before the slight movement behind him.

Will turned slowly. Salk stood in the doorway, half in shadow. The men stood looking at each other across the room. Salk spoke softly.

"Joanna's never been late before. Punctuality is one of her many—virtues."

For a moment, Will was anchored in the purity of his rage. His eyes burned at Salk. "She's not coming. Ever again. She won't be meeting you."

Salk lifted his eyebrows. "But she hasn't been meeting me."

Will's voice was hoarse. "Who?"

Salk half smiled. "Oh, I think you know." His eyes flicked around the room, came to rest on the bed. Will couldn't breathe.

"I told you that Joanna would pay any price to save Sydney's life. What is the highest price she could pay? What hell would she endure—again . . . over and over . . . for her child?"

Salk met Will's eyes, and Will suddenly saw all, as he'd imagined in tortured dreams: a child on the iron bed, shrinking back in terror, pressing herself against the wall at the sound of stumbling footsteps in the hall outside, sobs tearing at her throat as the door creaked open and her drunken father loomed in the doorway . . . burly, thick, swaying, his features coarse with anticipation . . .

"No . . . ," Will ground out in agony, his arms outstretched toward the empty bed. The charred walls seemed to vibrate with a lifetime of screams.

Salk spoke softly behind him. "Terrible. And yet—she was perfectly willing to pay. It is endlessly fascinating to me . . . what human beings will do for love."

Through the madness, Will grasped at a logic that was just out of reach. *Love. For love.*

"You don't own her," he said thickly. "She didn't sell anyone but herself."

Salk's smile flickered. "You caught that instantly, didn't you? I've always admired your legal skills." He spread his hands in concession. "True. It's no sin to sacrifice oneself for a loved one. The opposite, really, wouldn't you say? Her soul is her own." He shrugged dismissively. "What use is a soul, to me? My specialty—is suffering."

There was silence . . . just wind . . . and the crackling of some distant fire.

The dark man turned slightly toward the sagging bed, and his eyes seemed to caress it. "And have no doubt. Joanna's pain is exquisite. But she made a deal to save Sydney and she's keeping it . . . *every night*," Salk whispered.

Will lunged forward with an inarticulate cry of fury, but found he could not move more than a step. Salk's eyes gleamed in the dark. His tall silhouette seemed to waver like a shadow as the horror came for Will in waves.

"A pity you don't believe in God, isn't it?" Salk leaned against the door frame reflectively. "That's rumored to be of help in these situations." He glanced toward the bed again, and the planes of his face sharpened. "Unless you know God as Joanna and I do. Lovely Joanna knows from experience . . . all the prayers in the world, those long nights . . . and God never did a single thing to help her."

Will felt a bolt of molten agony as the child's weeping ripped through his body, her misery and pain and debasement and shame. He shuddered with her, his throat torn with gagging and sobs.

Joanna . . .

His eyes flew open. The bed was empty. It was only Will and Salk in the darkness, the wind moaning outside.

"Let her go," Will choked out.

Salk shrugged. "I could, it's true. But let's be clear. Sydney will die. If we return to the natural order of things."

Will stared across the room at Salk . . . through the suffocating finality of his words:

"The choice is yours," Salk said. "Your daughter—or your wife." A shutter creaked in the rising wind, like the ticking of a clock. "Joanna—or Sydney? Sydney—or Joanna?"

Will stood in the dark, his body aching, his face wet.

"Neither," he whispered.

Salk was still . . . his eyes burned like embers through the shadows.

"You never wanted Joanna. You don't care about Sydney." Will lifted his head. "It's me you want."

Salk smiled slowly in acknowledgment. He looked genuinely pleased. "I did come to you first, you know."

Will closed his eyes briefly, continued, "You want me in this election. You know I'll win. And the next . . ."

Salk raised a hand—a wave of dismissal. "Oh, our mutual friend Flynn is right. You're going to be president. Shall I tell you when?"

Will's voice was raw. "And you expect me to trade the world— for my wife and daughter." The two faced off across the tiny room. Will thought he could hear the muted sound of screams.

A smile played on Salk's lips, and something not quite human rippled over his features. "Do you know . . . in all my years . . . I've never had to make an offer. Invariably, people name their own price. Human beings always seem to know—exactly—their price."

Will turned his eyes away. His voice was heavy with effort. "I would do anything."

In the silence, Salk's voice was suddenly gentle, almost compassionate.

"Yes. I know."

Will straightened his shoulders. "Joanna and Sydney—free. You never come near them again."

Salk gestured, magnanimous. "You have my word. And in return . . . a small favor. One day." Again, that small, secret smile. "I'm certain something will come up."

He extended his hand for Will to shake. Will looked at his hand, then locked his gaze. Salk's eyes were bone-chilling.

Will reached—and clasped his hand.

CHAPTER FIFTY

In the early hours of dawn, the grandfather clock ticked in the hall.

The tortoiseshell cat slept curled on the straight-backed chair. The clock began to chime, and the cat raised its head, looked toward the front door. It opened almost without sound. Will walked in, closed the door behind him.

He stood in the hall for a long moment, the mirror reflecting him in the dark. He did not look into the glass as he walked to the stairs and up.

Upstairs, he paused in the hall to look into Sydney's room. She lay sleeping peacefully in bed, blond hair spread on the pillow, surrounded by stuffed rabbits, her face rosy with life.

In the master bedroom, the first silver streaks of dawn gleamed outside the window.

Joanna stirred on the bed . . . waking. She opened her eyes—to find Will was not beside her.

She sat up, startled . . .

. . . and saw him sitting in an armchair by the bed. Her face slowly smoothed out.

Will looked at her, drinking her in, his heart aching, for a moment unable to move. She was lovely as ever, but different than he'd ever seen her. It took him a moment to realize what it was: The shadows were gone from her eyes.

She frowned slightly, puzzled, as if something were new for her as well. She spoke impulsively.

"I feel . . . wonderful. Like—something incredible happened. But I can't remember."

Her face and her voice were so joyous—so *light*—Will suddenly understood that Salk had more than kept his promise: He'd wiped her memories completely clean. The pain, her pain, that had been there always, under the surface . . . all gone. And he felt a fierce moment of victory.

He spoke carefully, to keep his voice from shaking. "It was all a dream."

As he reached out to gather her in his arms, he caught a glimpse of himself in the dressing table mirror . . . and saw the darkness in his own eyes.

EPILOGUE

*The sound of muted cheers . . . of a marching band playing victo-
riously from far away. The sounds become louder . . .*

*In a packed convention hall, an audience rises in a thunderous ova-
tion as balloons drop from the ceiling like snow.*

*Will stands on a platform draped in red, white, and blue, with Syd-
ney and Joanna beside him. Brilliant camera flashes pop through the
hall.*

*Sydney is radiant, healthy; Joanna, beautiful and serene . . . the
crowd is cheering for Will, for them. Will's miked voice cuts through
the cheers . . . and the crowd quiets.*

"I thank you . . . and most of all . . . I thank my family."

*The crowd goes wild again as Will holds out a hand to Joanna.
Joanna tightens her arm around Sydney; she grasps Will's hand.*

*"These are challenging times. We have been awakened to dark
forces—at home and abroad. But I take the lessons I have learned
from my wife and daughter . . . and I have hope."*

In a lobby of the hospital, the television with Will's image
hangs from the corner of the wall. Will leans over the podium and
speaks to the crowd, to the camera.

"I have hope—that we will conquer our fears . . . and find strength from each other."

Salk stands in the lobby, watching the television, taking in Will's words, expressionless.

"I have hope—that by facing demons without and within—we will win our battles with the darkness that preys on the weak and on the innocent. I have hope that one day we will stand together, one family, united and whole . . . fearless in the face of our enemies."

Will stares out of the TV, directly at Salk.

"I have hope."

Salk turns away . . . and walks through the hospital lobby. As he walks, he cocks his head, listening to a sound, barely audible at first . . . but increasing . . .

The sound builds around him . . . the prayers of relatives keeping vigil . . . pleas in all languages . . . overlapping . . . rising and falling in waves . . . through anger, through tears.

Salk walks on, into the endless corridors of the hospital.

ACKNOWLEDGMENTS

Always, first, to Kimball Greenough, for his extraordinary contributions to this story and book.

My fabulous and much-loved agents, Scott Miller and Frank Wuliger, and to Sarah Self and Brian Lutz for their fine representation and help.

My spectacular editor, Marc Resnick, who makes me glad every day that I decided to try this novel thing, and the lovely and talented Sarah Lumnah, for her help and support.

Ben Sevier, for saying yes.

Again, Marc and Sarah, and Sally Richardson, Matthew Shear, Harriett Seltzer, Ellis Trevor, Talia Ross, Matt Baldacci, and the entire St. Martin's team.

Dr. Anton Nazaroff and Dr. Nicholas Saenz, for their medical expertise. The mistakes I've made and liberties I've taken are mine alone.

Ira Levin, for one perfect and perfectly terrifying story after another.

F. Paul Wilson, for worlds of inspiration . . . and major 7ths.

Ramsey Campbell, who makes the irrational rational, and vice versa.

Tim Lebbon—all that talent and that accent, too!

Doris Ann Norris, the two-thousand-year-old librarian, patron saint of mystery writers, and certainly the patron saint of my debut year.

Marc Evans, Laura Ziskin, Rachel O'Connor, and Vincent Newman for their creative contributions.

My awesome webmistress, Beth Tindall; and Michael Miller, Sheila English, and Adam Auerbach for their art.

The women of MUSE: Sarah Langan, Sarah Pinborough, and Deborah Le Blanc—true muses, all! www.MUSEfour.com

The extraordinary writers of Great Expectations: Jess Winfield, Franz Metcalf, and Elaine Sokoloff, for critique, support, and about a hundred lifetimes of friendship.

PJ Nunn, Katherine Nunn, and the dedicated crew at Breakthrough Promotions.

The whole gang at Murderati.com, for teaching me the business every day.

Sid Stebel and the Westside Writers Group for starting it all off.

Bob Levinson and the Killer Thriller Band—for a whole other dream.

The Slush Pile Players—ditto!

Heather Graham and Harley Jane Kozak: soul sisters in writing, music, theater, and life.

The Coven, again, because.

Katherine Fugate, for teaching me something about mothers and love.

Shireen Strooker, who made me go to dark places to find truth and beauty.

All you Berkeley people . . . official and honorary . . .

The authors, officers, and staffs of Sisters in Crime, International Thriller Writers, Mystery Writers of America, Horror Writers of

America, and Romance Writers of America, for creating these incredible communities.

And Nydia Ruscio, the moon-faced Buddha, who inspired this book with her short and courageous life.

Read on for an excerpt from the next book by
Alexandra Sokoloff

THE UNSEEN

Coming soon in hardcover
from St. Martin's Press

Outside the University Club, Laurel walked blindly on the oak-lined paths into the deepening sunset, the dark spiked silhouettes of Gothic buildings around her, with no idea where she was going. A dry wind whispered through the towering trees above, swirled leaves at her feet.

Great . . . that's just great. You can't even make it through a cocktail party. How do you think you're going to survive the year? she berated herself.

What am I doing here? What have I done? I've torpedoed my entire life, I've landed in the middle of nowhere, and I'm going to be fired if I don't come up with some knockout proposal by . . . I don't even know by when.

She laughed, but it was a strangled sound; she was again dangerously close to tears. Brendan Cody's questions taunted her, like hives prickling under her skin: *"And what about your hidden gifts, Professor MacDonald? Is this what you wanted to do when you grew up?"*

She forced herself to breathe, to look around. There was a medieval-looking archway in front of her, and she realized she was coming up on the Page Theater building, with its turrets and heraldic crests. Instead of heading back to her car, she'd started like a homing pigeon for the main yard of West Campus, in the direction of the Psych building.

Laurel turned on the path and looked up at Perkins Library, a looming shadow against the reddening sky. Suddenly she longed for the comfort of books around her. No matter how bad she felt, a library could always make her feel better.

She walked up the granite steps, and pushed through the fortress-like doors.

The entry of Perkins was a soaring three stories of gray stone, with a domed ceiling and arched windows that in the sunset streamed pink light into the column of space—a medieval chapel of a building. Even chattering students hushed when they walked through the heavy wooden doors. Laurel could feel all the molecules in her body rearranging into something peaceful and serene as she stepped through the gates and entered the sanctum.

She breathed in . . . and moved into an inner hall which led to a set of double doors of dark walnut with a brass plaque: SPECIAL COLLECTIONS LIBRARY.

The Library was a long, dark-paneled study with a fireplace at one end and heavy needlepoint drapes at the windows, groupings of antique sofas and divans, and recessed shelves—a cocoon of intimacy and concentration. Four other smaller rooms branched out from the first, each entirely lined with glass-paned bookshelves housing gorgeous volumes with gilt lettering on hand-tooled leather.

The first day Laurel found Special Collections, she couldn't believe that it wasn't packed with students, assistants, and professors, clamoring for any available space. She now suspected the emptiness had more than a little to do with the Special Collections librarian, Dr. Ward, a stout, quietly terrifying force of a woman who presided from a roll-top desk near the front door.

Ward wore thick round glasses that made her look vaguely like an enraged owl and black hair cut in a severe pageboy. Her gaze could cut a student down at twenty paces.

On her first day Laurel had approached the desk with no small amount of trepidation. "I'm Laurel MacDonald. I'm a new professor in the Psych Department."

Dr. Ward had looked at her unblinkingly through Coke-bottle lenses, without speaking.

Laurel had summoned her courage. "I wondered if study time in the reading room had to be reserved? I'd like to sign up for a regular time, if that's possible."

The librarian had looked at her without smiling, and answered, "No reservation required. Help yourself to a table."

Laurel had glanced around the magnificent, empty room. "But . . ."

Dr. Ward had looked up at her, waiting.

Laurel had half-laughed. "I just can't believe there aren't more people here."

"That would require taking the initiative of asking," the librarian replied dryly.

"Oh. I see." Laurel said, unable to believe her luck.

After that, almost daily, she'd brought her notes and journals and texts with her and found a table in one of the small, leather volume–lined alcoves off the main room, where she could immerse herself in preparation for a lecture or let her mind wander to possible research topics. Dr. Ward always nodded to her—the briefest of nods, it was true, but even that was somehow comforting, the beginning of familiarity and routine, when everything else around her was so uncertain. And under that, perhaps, was even a hint of adventure.

Of course Laurel had immediately had a favorite room, which was probably why she missed the display case in her first few days of haunting the rare books room.

But tonight she nodded to Ward—who was completely alone at her command post—and just wandered through the rooms, actually looking at them for the first time. Beyond the superb collections of hand-bound books, each room contained large glass exhibit cases displaying parts of the collection: fine sketchbooks open to drawings of native birds, maps documenting Blackbeard's journeys on the Carolina coast, architectural renderings of the campus buildings. Laurel moved into another, smaller alcove room. Like the other rooms it had a large glass exhibit case, this one on four legs, table-like. She drifted over to the exhibit, expecting more sketchbooks.

What she saw instead froze her in her tracks.

She was looking down at a set of five cards of playing-card size, glossy and with a grayish tinge, as if they were of some age. There was something both archaic and strikingly familiar about the bold black symbols: a circle, a cross, three wavy lines, a square, and a star.

She *knew* those symbols.

Even as Laurel began to read the explanatory placards, a rush of connections and memories was flooding into her consciousness. The cards with their strange symbols were Zener cards, designed for use in the groundbreaking ESP tests of Dr. J.B. Rhine.

The Duke University parapsychology lab was founded in 1927 by Dr. William MacDougall and Dr. Joseph Banks Rhine, with the mission of searching for scientifically quantifiable evidence of paranormal, or psi phenomena, such as ESP and psychokinesis.

Laurel was experiencing a rush of emotions she was having trouble identifying. *I didn't remember any of that*, she realized with an uneasy feeling. *Not the Zener cards, not that the Rhine*

*lab was at Duke, not any of it. I came straight to the Rhine lab—
without even knowing what I was doing.*

The coincidence was unsettling . . . and thrilling.

Laurel turned back to the glass case and quickly read
through the rest of the historical placards of the display:

**The Duke laboratory was the most famous parapsycholo-
gy lab in the U.S. Over the lab's thirty-eight-year history,
Dr. Rhine and his researchers scientifically proved the
existence of ESP, using "forced choice" testing with the
Zener cards, in which a test subject would attempt to
guess each card turned up from a deck of twenty-five
cards, consisting of five sets of five simple symbols. Pure
statistical chance would be a guess of five cards out of
twenty-five, or 20% correct. Any score significantly
greater than 20% correct was an indicator of psychic ability, and Rhine discovered test subjects who could predict
the cards with accuracy far beyond statistical chance.**

**Rhine went on to perform laboratory tests of psychokinesis,
the movement of objects by the mind, using motorized dice-
throwing machines, and in the early sixties his researchers
conducted field investigations of poltergeists—**

Poltergeists?? Laurel thought, startled. *They were investigating poltergeists?*

And then on the last placard she found the sentence that
stopped her dead.

**When the Duke lab closed in 1965, seven hundred boxes of
research material from the parapsychology lab were sealed
and stored in the basement of Perkins Library. Now for the**

first time in forty-five years, those boxes have been made available for public viewing.

Laurel had to read that last part three times before it fully sank in.

Seven hundred boxes of original parapsychology research material? Right here in this very building? And available to anyone who wanted to look?

She felt behind her for a chair and sat down hard, a little breathless.

It was sensational. It was unbelievable, really, that she hadn't read or heard about any of this.

Surely someone had already claimed this topic, was writing articles, papers . . .

J. Walter Kornbluth's voice suddenly spoke clearly in her mind. *"These days nothing less than a book will do."*

Laurel looked over at the glass case, and thought with crystal clarity: *This is my book.* Her whole body was tingling, her face warm and flushed.

"Whoa, whoa, whoa. What are you thinking?" she muttered to herself. *You're a psychologist, a research professor. Parapsychology has nothing to do with your life's work.*

Oh, yeah? What life's work is that? some alien voice whispered back, mocking.

And she hadn't felt a rush like this, hadn't felt so enthused, since, well, since before *the dream* had shattered all her ideas of reality.

She stood from the chair, moved past the glass case in a daze, and went into the main reading room to find the librarian.

As always, Laurel approached Dr. Ward with some caution, and hovered some distance from the tall dark desk. Ward looked her over with a slightly raised eyebrow, and Laurel remembered that she was still in cocktail-party attire. She felt blood rising to her cheeks, and surreptitiously tugged the hem of her dress down, but she pushed on, undaunted.

"I was just reading the exhibit about the Rhine Lab," Laurel began. "I guess I somehow forgot that all that happened here." Now that she was thinking about it, it hit her that the auditorium downstairs in the Psych building was called Zener Auditorium.

I must have been asleep not to put it together.

"Thirty-seven years," Ward agreed laconically. "Put the University on the map."

Encouraged by a whole two sentences from the librarian, Laurel pressed on.

"I'd like to know more about it—all. Is it really true that there are seven hundred boxes of original research files right here in the library?"

"There are indeed," the librarian said, without smiling. Laurel had never seen her smile. "Seven hundred boxes."

"And anyone can just look at them—anytime?"

"Anytime."

Laurel hazarded another question. "Is there some study being done, then?"

"A study?" the librarian repeated.

"A research project, or a book being written, or . . . I mean . . . if seven hundred boxes of original research have just been opened to the public, isn't someone going through them?"

"There have been a few," the librarian said noncommittally.

"Of course," Laurel said. There was no reason to think otherwise, after all.

She stood for a moment, then suddenly asked. "Do you know why the department was shut down?"

The librarian raised her eyebrows.

"I mean, the parapsychology lab was so famous, and then . . ." She nodded back toward the display case. "The placards say it shut down entirely in nineteen sixty-five . . . and I wondered why."

"Dr. Rhine was approaching retirement age and wanted to move his research to a private institution where he could continue his work," Dr. Ward recited without inflection.

"That makes sense," Laurel admitted. "But why shut the whole lab down?"

Again the librarian looked up at her without speaking. This time Laurel didn't notice; she was off on her own train of thought.

"And *why* were they sealed?"

The librarian regarded her impassively.

"The boxes," Laurel elaborated, for some reason feeling uneasy. "Why were they sealed, for all those years?"

"I couldn't tell you," Ward said.

"Thanks," Laurel said. Her mind was already a million miles away. "Thanks."

She walked out of Special Collections, out of the library, moving in a daze through Gothic stone arches, her face lit from within.

Seven hundred boxes.

What in the world might be in them?